T0220479

Diamond Electronics and Bioelectronics — Fundamentals to Applications III

MATERIALS RESEARCH SOCIETY
SYMPOSIUM PROCEEDINGS VOLUME 1203

Diamond Electronics and Bioelectronics — Fundamentals to Applications III

Symposium held November 30 – December 3, 2009, Boston, Massachusetts, U.S.A.

EDITORS:

Philippe Bergonzo

CEA-LIST
Saclay, France

James E. Butler

U.S. Naval Research Laboratory
Washington, D.C., U.S.A.

Richard B. Jackman

University College London
London, UK

Kian Ping Loh

National University of Singapore
Kent Ridge, Singapore

Milos Nesladek

Hasselt University
Hasselt, Belgium

Materials Research Society
Warrendale, Pennsylvania

CAMBRIDGE UNIVERSITY PRESS
Cambridge, New York, Melbourne, Madrid, Cape Town,
Singapore, São Paulo, Delhi, Mexico City

Cambridge University Press
32 Avenue of the Americas, New York NY 10013-2473, USA

Published in the United States of America by Cambridge University Press, New York

www.cambridge.org
Information on this title: www.cambridge.org/9781107408111

Materials Research Society
506 Keystone Drive, Warrendale, PA 15086
http://www.mrs.org

© Materials Research Society 2010

This publication is in copyright. Subject to statutory exception
and to the provisions of relevant collective licensing agreements,
no reproduction of any part may take place without the written
permission of Cambridge University Press.

This publication has been registered with Copyright Clearance Center, Inc.
For further information please contact the Copyright Clearance Center,
Salem, Massachusetts.

First published 2010
First paperback edition 2012

Single article reprints from this publication are available through
University Microfilms Inc., 300 North Zeeb Road, Ann Arbor, MI 48106

CODEN: MRSPDH

ISBN 978-1-107-40811-1 Paperback

Cambridge University Press has no responsibility for the persistence or
accuracy of URLs for external or third-party internet websites referred to in
this publication, and does not guarantee that any content on such websites is,
or will remain, accurate or appropriate.

CONTENTS

* Invited paper

NANODIAMOND

* Invited paper

GRAPHENE

DOPING

SURFACE MODIFICATIONS

CHARACTERIZATION

* Invited paper

DEVICE APPLICATIONS

(1) Bioapplications

DEVICE APPLICATIONS

(2) Electrochemistry

* Invited paper

DEVICE APPLICATIONS

(3) Radiation Detectors

DEVICE APPLICATIONS

(4) Field Emission

DEVICE APPLICATIONS

(5) Electronics Applications

DEVICE APPLICATIONS

(6) Optical Applications

* Invited paper

PREFACE

Symposium J "Diamond Electronics and Bioelectronics — Fundamentals to Applications III," held November 30–December 3 at the 2009 MRS Fall Meeting in Boston, Massachusetts, followed the first two MRS symposia dedicated to diamond technology, held in Fall 2006 and 2007. It was the largest yet, with over 140 papers being presented. Sessions were dedicated to Diamond Nanotechnology, Quantum Diamond, Diamond Nanoscale Sensors, Thermal and Mechanical Applications of Diamond, Progress in CVD Diamond Growth, Diamond Electrochemistry, Nanodiamond and Nanowires, Low-dimensional Carbons, Diamond Bioelectrochemistry, Diamond Biosensing, Diamond Industrial Technologies and Applications, Diamond Single Crystal and Defects, Diamond Doping and Transport, Diamond Surface Devices, Diamond for Thermionic Conversion and Electron Emission and Diamond Devices. Sessions were extremely well attended and the audience participation high.

Picking highlights is a notoriously difficult business, but the presentation made by Deleonibus (CEA-LETI, Grenoble, France), 'Nanoelectronics Roadmap and the Opportunities for Carbon Based Electronics in Diversification' opened the meeting with a thought provoking discussion of how diamond may play an important role in future ULSI technology. Quantum information processing is looking increasingly towards diamond, with the fact that single photon devices based on diamond are becoming commercially available being highlighted (Aharonovich, University of Melbourne, Australia). It was noticeable how the use of nanodiamonds had grown in importance since earlier meetings (for example, 'Nanodiamond Particles in Electronics and Optical Applications,' Shenderova, ITC, Raleigh, North Carolina, USA). Diamond Biotechnology also increased in importance, with the contribution from Yang from IAF (Germany) highlighting key approaches for the fabrication of diamond biosensors. The surface electronic properties of graphene were interestingly contrasted with diamond by Wee (NUS, Singapore). With a view to future energy needs, Nemanich (Arizona State University, USA) discussed 'Thermionic Emission and Energy Conversion Based on Diamond.'

The organizers would like to thank the staff and officers of MRS for their very efficient handling of all aspects of the symposium, which made our job much easier. We would also like to thank the authors within this volume for their timely submission of some very interesting papers. The symposium benefited enormously from the generosity of our sponsors:

> Advanced Diamond Technologies Inc
> AGD Material CO, Ltd
> Apollo Diamond
> Applied Diamond Inc
> CEA-LIST, Saclay France
> Element Six
> Group4Labs
> Lambda Technologies

SEKI Technotron USA
sp3 Diamond Technologies

who are all very gratefully acknowledged.

"Diamond Electronics and Bioelectronics — Fundamentals to Applications IV" will be held as Symposium A during the 2010 MRS Fall Meeting (November 29–December 3, 2010). We hope to see you there.

Milos Nesladek
Philippe Bergonzo
James E. Butler
Richard B. Jackman
Kian Ping Loh

June 2010

MATERIALS RESEARCH SOCIETY SYMPOSIUM PROCEEDINGS

MATERIALS RESEARCH SOCIETY SYMPOSIUM PROCEEDINGS

Prior Materials Research Society Symposium Proceedings available by contacting Materials Research Society

Growth

Mater. Res. Soc. Symp. Proc. Vol. 1203 © 2010 Materials Research Society   1203-J17-25

Compact and Efficient HFCVD for Electronic Grade Diamonds and Related Materials

R. D. Vispute*, Andrew Seiser, Geun Lee, Jaurette Dozier, J. Feldman, L. Robinson, B. Zayac, and A. Grobicki

Blue Wave Semiconductors, Inc., 1450 S. Rolling Rd, UMBC Tech Center, Baltimore, MD 21227

*contact: rd@bluewavesemi.com

ABSTRACT

A compact and efficient hot filament chemical vapor deposition system has been designed for growing electronic-grade diamond and related materials. We report here the effect of substrate rotation on quality and uniformity of HFCVD diamond films on 2" wafers, using two to three filaments with power ranging from 500 to 600 Watt. Diamond films have been characterized using x-ray diffraction, Raman Spectroscopy, scanning electron microscopy and atomic force microscopy. Our results indicate that substrate rotation not only yields uniform films across the wafer, but crystallites grow larger than without sample rotation. Well-faceted microcrystals are observed for wafers rotated at 10 rpm. We also find that the Raman spectrum taken from various locations indicate no compositional variation in the diamond film and no significant Raman shift associated with intrinsic stresses. Results are discussed in the context of growth uniformity of diamond film to improve deposition efficiency for wafer-based electronic applications.

INTRODUCTION

Diamond films are attractive for electronic, optical, and mechanical applications because of their excellent hardness, strength, chemical and thermal stability, thermal conductivity, breakdown voltage, hole and electron mobility, radiation hardness, and optical transmission [1-3]. Enormous efforts have been made in studying diamond thin film growth by several chemical vapor deposition techniques including plasma assisted CVD (microwave, RF, DC), hot filament, electron and laser assisted, and ion beam assisted [4-13]. Large area and uniform diamond films are required for wafer-based semiconductor device applications including Microelectromechanical (MEMS) devices. Among many chemical vapor deposition techniques, HFCVD is the simplest and most versatile technique for growing diamond films on various substrates over a large area. Owing to its low capital cost and the ease of scalability, HFCVD has been considered as a very useful technique for producing low cost diamond films for industrial applications. In order to exploit HFCVD diamond for wafer-based electronics applications such as cold cathode field emitters, high thermal conductivity substrate materials for white LEDs, and high power Schottky diodes, diamond deposited substrate areas need to be scaled up without loss of uniformity or film quality.

The main motivation of the present study is to identify the effects of various processing parameters on growth of diamond films on a rotating substrate or wafer to achieve uniform diamond films over large area substrates while minimizing thermal and power requirements for efficient reactor operation. In this context, we have developed a commercial HFCVD reactor in which the substrate or wafer can be rotated at 10 rpm while keeping the number of filaments to be as low as two or three. Our research efforts allowed us to develop a low cost, highly efficient, and high throughput commercial HFCVD system for production of high quality diamond films. Here, we discuss optimization of process parameters on the growth of diamond films with emphasis on quality, morphology, and uniformity of diamond films. Results are discussed in the context of enhancing crystallite size, which is useful for various applications of diamond films in electronics.

EXPERIMENTAL

The stainless steel, double-wall, water cooled vacuum chamber was design to accept a 2" dia wafer (capable up to 4" dia wafer) through a load lock chamber (Fig. 1). A load lock transfer chamber is attached to the gate wall and it allows for easy and quick access to the substrate without breaking vacuum in the main HFCVD chamber. A multi-axis positioner is connected to the load lock and allows the transfer arm to be tilted in the desired position. The load lock chamber is also beneficial for carburization of filaments before loading wafer for deposition. The bottom port holds a flange with a rotating substrate heater capable of reaching 850 °C with linear z directional travel of 2 inches that allows variable distance between substrate and filaments. The components of the filament assembly are presented in Fig. 1. It shows a filament assembly capable of having a number of filaments (0.5 mm dia and 8 cm

3

long) from 2 to 8 with spring loaded moly holders to adjust tension in the filaments associated with change in length due to thermal expansion and sagging of the filaments. Tungsten filaments were placed 15 mm apart with one position originating near the center of a 2" substrate. The substrate heater, measuring 2.2" in diameter suitable for a 2" substrate, is mounted on a vertical z stage with susceptor capable of rotating at 10 rpm during diamond deposition. The purpose of substrate rotation is to introduce uniform diamond deposition throughout wafers with diameter equal or larger than 2". For optimization studies, we mainly focused on 2" single side polished silicon wafers. Prior to diamond growth, wafers were first cleaned in an ultrasonic bath

Fig. 1 Blue Wave HFCVD System with filament arrangement capable of reaching more than 2000°C for growth of diamond films.

of acetone for 5 min. and then blown dry by compressed air. The substrates are then seeded with either 0-0.2 micron or 3 micron diamond pastes by applying the paste mechanically onto the substrate surface followed by ultrasonic bath treatment in diamond slurry for 15 min. and then finally rinsed and then cleaned in ultrasonic bath of acetone and methanol.

To avoid poisoning of tungsten impurities and enhance chemical purity in the deposited thin films, tungsten filaments were pretreated and carburized for 30 min. to 1hr. New tungsten filaments were installed every experiment to ensure reproducibility of deposition process. The filaments were exposed to higher methane concentration (10%-50%) at approximately 1500°C for 10min, then at 1800–1900°C for 5min and finally for a brief moment at 2100°C to form a thin tungsten carbide layer, which prevents tungsten evaporation during subsequent growth. This filament pretreatment is similar to that of reported by Yap et.al. [14] which confirms no sign of W contamination in their samples as verified by X-ray photoemission spectroscopy. Once the carburization of the filaments is completed, the pretreated wafer is transferred to growth chamber without breaking vacuum and the process parameters for diamond growth are set. Filament temperature was monitored using an infrared pyrometer to 2000°C. Typical power for the filaments was found to be about 500 watt to reach deposition of good quality diamond films. The distance between filament and substrate was kept constant at about 10 mm. HFCVD diamond samples were characterized by scanning electron microscopy, Raman spectroscopy, and x-ray diffraction, and electrical resistivity measurements. Raman spectroscopy was performed using confocal Raman Imaging mode (alpha 300 R) made by WITech. Raman spectra are collected with a high sensitivity confocal microscope connected to a high-throughput spectrometer equipped with a CCD camera. Excitation laser wavelength used for the Raman spectroscopy was 532nm.

RESULTS AND DISCUSSION

Initially, we optimized the diamond deposition process using conventional approaches to deduce the effect of wafer rotation on diamond quality. For optimization, we systematically varied methane to hydrogen ratio for growth of diamond films without substrate rotation. Fig. 2 shows scanning electron micrographs of the HFCVD diamond films grown at various CH_4/H_2 ratios.

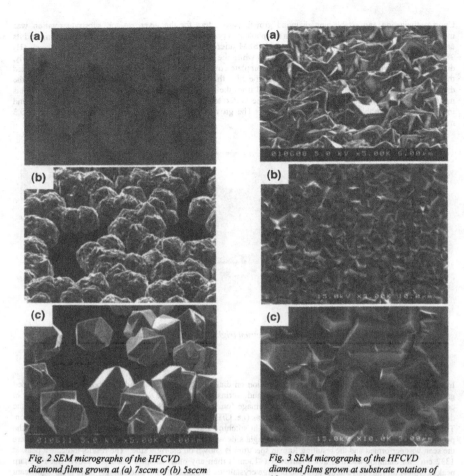

Fig. 2 SEM micrographs of the HFCVD diamond films grown at (a) 7sccm of (b) 5sccm and (c) 2 sccm of CH4 in H2 of about 93-98 sccm of H₂.

Fig. 3 SEM micrographs of the HFCVD diamond films grown at substrate rotation of 10 rpm for (a) 6 hrs and 25 hrs respectively. Fig 3 (c) is the magnified image of 3 (b).

Fig. 2(a) corresponds to diamond film grown at higher methane flow rate of 7 sccm (H₂ of 93sccm). From the SEM image it is clear that the growth of the film is supersaturated with carbon (typically observed sp² bonded material in sp³ matrix). As the methane content was reduced to 5 sccm (H₂ of 95 sccm), diamond crystallites began to show facets. Upon further decrease in methane flow to 1.5 to 2 sccm (H₂ 98sccm) highly faceted diamond crystals are visible in the microscope. Scanning electron micrograph of this sample is shown in Fig. 2(c). Diamond crystals of about size 4-5 micron are clearly visible in this micrograph. Thus, this study provided optimum growth conditions for the formation of polycrystalline diamond films irrespective of substrate rotation. This means that CH₄/H₂ ratio, and the filament temperature are adequate for diamond growth. Importantly, note that the film grown at lower methane conditions has poor surface coverage indicating poor lateral growth of crystallites or significantly less multiple nucleation of diamond crystallites.

Upon optimization and establishing diamond growth parameters for the given reactor, substrate rotation was introduced to study its effect on diamond film morphology, crystallite size of microcrystalline diamonds film, and its uniformity and quality. Figs 3 (a) and 3(b) show SEM micrographs of the diamond films grown at 2 sccm (H_2 98sccm) for the duration of about 6 hrs and 20 hrs while the wafer was rotated at 10 rpm, respectively. Clearly, diamond growth with high nucleation density and complete coverage is observed for the rotated substrate. We noted that diamond film coverage was within 5-7% of the film thickness over the 2" wafers. We also noted that the diamond crystals grow larger with time as observed from the SEM micrograph of the thicker film (Fig. 3c). This means the substrate rotation uniformly exposes the wafer to atomic hydrogen and carbon radical species and improves uniform growth of diamond crystallites. The growth rate under substrate rotation is found to be 0.3 micron/hr for the given power.

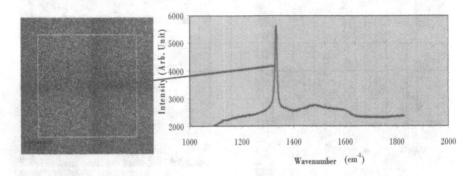

Fig. 4 Video micrograph (left) and Raman spectrum (right) of the typical HFCVD diamond film grown under substrate rotation.

In order to further study the effect of sample rotation on diamond quality (Sp^3 bonded carbon versus sp^2 bonded graphitic material), compositional homogeneity, and intrinsic stresses, we conducted Raman spectroscopy in confocal imaging mode. Fig, 4 shows the video image (on the left) and Raman spectrum (on the right) for the deposited diamond sample under optimum conditions (i.e. CH_4 of 2 sccm and H_2 of 98 sccm) with sample rotation. The data was recorded with the integrated white light stitching function of the microscope. The cross indicates the positions where the single spectrum shown on the right side was recorded. The yellow box indicates the area where the scan was performed and the resulting Raman spectrum is shown on the right side. The results indicate a peak at 1332 cm^{-1} for diamond and it is much narrower (<13cm^{-1}) than any other samples discussed above. Such a sharp peak is typical for microcrystalline diamond or large crystallites in diamond film. Interestingly there is no significant increase of the peak near 1500 cm^{-1} that is associated with graphitic or sp^2 bonded material. Thus substrate rotation does not adversely affect the quality of the diamond film.

We have also performed Raman scans at various locations on the diamond deposited wafer with substrate rotation as shown in Fig. 5. After performing a scan with the high resolution grating, the diamond line of each of the 22500 spectra (corresponding to each pixel of the 150x150 image) was fitted using a Lorenzian fit. The images below show the results of two of the fit parameters: the line width (FWHM) and the exact peak position of the line. The spectra beneath the images show the extreme cases color coded according to the images taken for diamond film near the center and near the edge of the wafer. The results indicate that the line width broadening of the Raman peak near the center of the wafer is within 5-8 cm^{-1} and peak position is within 2-3 cm^{-1}. Similar Raman results are found near edge of the wafer as shown in Fig. 5(b). From these results we conclude that the substrate rotation introduces uniform diamond deposition process without significant increase in sp^2 or graphite phase material or any significant stresses associated with compositional changes throughout the wafer.

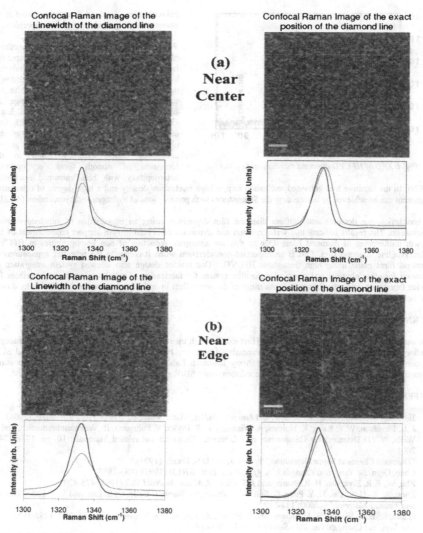

Fig. 5 *Confocal Raman images and corresponding Raman scans near (a) the center and (b) the edge of the diamond deposited wafer. Left side images and spectra indicate variation in line width of characteristic 1332 cm⁻¹ peak while right side images and spectra indicate variation in peak position over the wafer's various locations.*

We have also conducted x-ray diffraction studies on the optimally grown diamond films. The films are found to be polycrystalline as expected from SEM results. However, major orientation of the diamond microcrystals is found to be {111} normal to the substrate plane. This is consistent with the surface morphology obtained during HFCVD that depends critically upon the gas mixing ratio and the substrate temperature [3]. The electrical resistivities of these

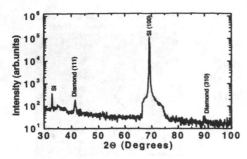

Fig. 6 XRD of HFCVD diamond thin film on Si (100).

polycrystalline diamond films were found to be greater than 10^{10} ohm-cm at room temperature.

We are studying additional controllable methods of enhancing nucleation such as bias induced ion bombardment. Our reactor design is capable of biasing substrates up to -250V. Simply adding a negative bias of a few hundred volts to the substrate will allow the ions to damage the surface and implant into the lattice which are known to favorably affect diamond growth. One can envisage (at least) two routes to this objective: one has to either identify growth conditions which naturally result in the formation of smooth films, or optimize heteroepitaxy with bias enhanced nucleation (BEN). In the negative bias enhanced nucleation step, a high nucleation density and a high degree of orientation alignment can be achieved by bombarding the Si substrates with positive ions of hydrogen and hydrocarbon [4-13].

In conclusion, we demonstrated uniform diamond film deposition using an introduction of substrate rotation mechanism. The results indicate that wafer rotation and dynamics of crystal growth support enhancement of the growth uniformities of diamond films. Due to this we anticipate reduction of non homogeneities in HFCVD polycrystalline diamond film. This is an important consideration when it comes to the potential applications of diamond films made using high throughput HFCVD. Our reactor design also allowed system integration of diamond for other physical and chemical deposition system for fabrication of metal-semiconductor interfaces for device fabrication, and other common applications of diamond films in optical and power devices (Schottky diodes and field emitters).

ACKNOWLEDGEMENTS

The authors thank Thomas Dieing and Robert Hirche of WITech Instruments, for helping with Raman Spectroscopy. Authors also thank Prof. Jimmy Davidson, Vanderbilt University Nashville TN for helpful discussion and SEM support. Authors also thank Dr. Ken Jones, Army Research Laboratory, Adelphi MD for helpful discussion. Financial support from the National Science Foundation under SBIR grant 0823126 is gratefully acknowledged.

REFERENCES

1. Jie Yang, Weixiao Huang, T.P. Chow, and James E. Butler, Mater. Res. Soc. Symp. Proc. Vol. 905E, (2006).
2. J. L. Davidson, W. P. Kang, K. Holmes, A. Wisitsora-at, P. Taylor, V. Pulugurta, R. Venkatasubramanian, and F. Wells, "CVD Diamond for Components and Emitters," Diamond and related Materials, 10, pp. 1736-1742, 2001.
3. "Diamond Chemical Vapor Deposition" by H. Liu and D.S. Dandy (1995).
4. Chen, Qijin, Jie Yang, and Zhangda Lin, *Appl. Phys. Lett.* 67(13) (1995) 1853-1855.
5. Zhu, W., F. R. Sivazlian, B. R. Stoner, and J. T. Glass, *J. Mater. ResNucl* 10.2 (1995) 425-430.
6. Zhou, X. T., H. L. Lai, H. Y. Peng, C. Sun, W. J. Zhang, N. Wang, I. Bello, C. S. Lee, and S. T. Lee, *Diamond and Related Materials* 9 (2000) 134-139.
7. Gupta, S., B. R. Weiner, W. H. Nelson, and G. Morell, *Journal of Raman Spectroscopy* 34 (2003) 192-198.
8. Liu, Wei, and Changzhi Gu, *Thin Solid Films* 467 (2004) 4-9.
9. Wang, Q., C. Z. Gu, Z. Xu, J. J. Li, Z. L. Wang, X. D. Bai, and Z. Cui, *J. Appl. Phys.* 100 (2006) 1-5.
10. Kromka, A., F. Balon, T. Danis, J. Pavlov, J. Janik, V. Dubravcova, and I. Cerven, *Surface Engineering* 19.6 (2003) 417-420.
11. Janischowsky, K., W. Ebert, and E. Kohn, 12 (2003) 336-339.
12. Larijani, M. M., A. Navinrooz, and F. Le Normand, *Thin Solid Films* 501 (2006) 206-210.
13. Feng, Xu, Zuo Dunwen, Lu Wenzhuang, and Wang Min, *Transactions of Nanjing University of Aeronautics & Astronautics* 24.4 (2007) 317-322.
14. H.Y. Yap, B. Ramaker, A.V. Sumant, R.W. Carpick, Diamond & Related Materials 15, 1622–1628 (2006).

Mater. Res. Soc. Symp. Proc. Vol. 1203 © 2010 Materials Research Society   1203-J17-15

Self-Assembly of Cylinder-Forming Block Copolymers on Ultrananocrystalline Diamond (UNCD) Thin Films for Lithographic Applications

Muruganathan Ramanathan[*†]; Seth B. Darling[†]; Anirudha V. Sumant[†]; Orlando Auciello[††]

†Center for Nanoscale Materials and ‡Materials Science Division, Argonne National Laboratory, 9700 S. Cass Ave., Argonne, IL 60439

ABSTRACT

Block copolymers (BCPs) consist of two or more chemically distinct and incompatible polymer chains (or blocks) covalently bonded. Due to the incompatibility and connectivity constraints between the two blocks, diblock copolymers spontaneously self-assemble into microphase-separated nanoscale domains that exhibit ordered 0, 1, 2 or 3 dimensional morphologies at equilibrium. Commonly observed microdomain morphologies in bulk samples are periodic arrangements of lamellae, cylinders, or spheres. Block copolymer lithography refers to the use of these ordered structures in the form of thin films as templates for patterning through selective etching or deposition. The self-assembly and domain orientation of block copolymers on a given substrate is critical to realize block copolymer lithography as a tool for large throughput nanolithography applications. In this work, we survey the morphology of cylinder-forming block copolymers by atomic force microscopy (AFM). Three kind of block copolymers were studied: a) poly(styrene-*block*-ferrocenyldimethylsilane), PS-*b*-PFS b) poly(styrene-*block*-methylmethacrylate), PS-*b*-PMMA and c) poly(styrene-*block*-dimethylsiloxane) PS-*b*-PDMS. Block copolymers were dissolved in a neutral solvent for both blocks (toluene) in order to obtain solutions of various concentrations (1 and 1.5 wt %). From these solutions, films were prepared by spin casting on ultrananocrystalline diamond (UNCD) thin film substrates. Results indicate that PS-*b*-PFS exhibits chemical and morphological compatibility to the UNCD surface in terms of wetting and domain control. A systematic comparison of self-assembly of these polymers on silicon nitride substrates demonstrates that UNCD thin films would require pre-treatment to be considered as a substrate for BCP lithography.

INTRODUCTION

The quest for developing materials at the nanoscale is escalating since early this decade. As the dimension of the materials reduced from the bulk to a system composed of few tens of atoms at the nanoscale, fascinating properties in magnetism, catalysis, electronics, optics, mechanics and ferroelectricity emerged. The transition from fundamental science to real world applications, for instance, in the fields of energy (photovoltaics), electronics (miniaturized ultrahigh capacity storage devices), developmental biology (therapeutics) and biomedical engineering, greatly relies on developing nanostructured materials. Among the various nanostructures, nanowires receive special attention due to their relevance in many technological applications. [1,2]

Both bottom-up and top-down methodologies have been successfully implemented to develop nanowires of various materials. [3-8] In this work we explore the possibility of using block copolymer lithography on UNCD substrates for patterning nanostructures. UNCD films are

---

* Corresponding author email: Nathan@anl.gov Phone: 630-252 7789 Fax: 630-252 4646

distinguished from microcrystalline diamond films (1-5 μm grain size) and nanocrystalline diamonds films (10-200 nm) by the smallest grain size (2-5 nm), lowest surface roughness (5-7 nm) and percentage of sp^3 bonding, as compared to the other diamond films mentioned above. [9, 10]

Block copolymer lithography is an emerging lithographic method for patterning materials at the nanoscale. For etch masking/lithographic applications, diblock copolymer thin films spontaneously form nanometer-scale patterns over a large area.[11-13] Furthermore, each block of the copolymer can be chosen for a specific application, such as to tailor substrate compatibility and high differential etch resistance. Selective processing of one block relative to the other is possible by use of chemical or physical dissimilarities between the two blocks.[14] Substrates are often preferentially wet by one of the blocks of the copolymer and, using the cylindrical phase as an example, the cylinders tend to orient parallel to the substrate. Strategies such as chemical modification and the application of external fields can be used to neutralize or overcome this tendency. Wide-scale implementation of BCP lithography in nanomanufacturing faces many challenges including pattern perfection over macroscopic areas, dimensional control of features within exacting tolerances and margins, and registration and overlay. These issues are being addressed using strategies like directed assembly of block copolymers on topographically patterned substrates,[15-17] or directed assembly of block copolymers on chemically nanopatterned substrates.[18-21]

The focus of this paper is to demonstrate the self-assembly characteristics of three BCPs (shown in scheme 1) a) poly(styrene-*block*-ferrocenyldimethylsilane), PS-*b*-PFS b) poly(styrene-*block*-methylmethacrylate), PS-*b*-PMMA and c) poly(styrene-*block*-dimethlysiloxane), PS-*b*-PDMS on UNCD thin films. Whereas all these polymers form cylindrical microdomains they are dissimilar in their chemical and physical (etch resistance) characteristics. Understanding their self-assembly and domain orientation control is necessary to identify a suitable polymer and to ascertain if any pre-treatment is required for its self-assembly on the UNCD surface to be used as an effective mask to develop large throughput, NEMS-relevant patterns. A systematic comparison of the obtained results on UNCD substrate to the results obtained on silicon nitride substrates provides further insight into the suitability of UNCD for nanopatterning via BCP lithography.

Scheme 1 chemical structures of a) PS-*b*-PFS b) PS-*b*-PDMS and c) PS-*b*-PMMA

EXPERIMENTAL DETAILS

Substrate pretreatment and UNCD growth:
UNCD films were grown in a DiamoTek 1800 series 915 MHz, 10 KW MPCVD system (Lambda Technologies, USA). A detailed description of this tool can be found in Ref.22.[22] To

initiate the nucleation, substrates were seeded in an ultrasonic bath using a nanodiamond suspension (from International technology Center, Raleigh, NC). Details of the seeding process are explained in more details in ref 23. [23] A nearly atomically smooth surface of UNCD is obtained by etching away the parent substrate by exposing the nucleation side of the UNCD film in contact with the Si substrate via chemical etching of the substrate. Details of chemical etching process and transferring free-standing UNCD on to the other dummy substrate is described elsewhere.[23]

Block copolymer thin film:

All BCPs used in this work are asymmetric, synthesized by anionic living polymerization and obtained from Polymer Source as custom synthesized BCPs.

PS-*b*-PFS with a molecular weight of 90,000 g mol^{-1} (M_n=60,000 PS/30,000 PFS) and a polydispersity index of 1.2 was used as received. Thin films were spin-cast at a spin speed of 4000 rpm from toluene solutions with concentration of 1 wt.% onto clean 5 mm^2 UNCD and silicon nitride substrates. Solvent annealing was carried out in a closed bell jar with a saturated toluene atmosphere (partial pressure of 3.4×10^3 Pa as calculated with the Antoine equation). As deposited, the thick glassy PS layer present at the surface makes imaging microphase separated domains difficult, therefore, prior to imaging the PS phase was removed using oxygen plasma reactive ion etching. Oxygen-RIE was performed in a March CS-1701 RIE system at a pressure of 100 mTorr, 20 W power, 7 sccm oxygen flow rate and for 3 min. etch time. AFM imaging was performed with a Veeco MultiMode V system equipped with active vibration isolation. Tapping-mode etched silicon probes were used for imaging.

PS-*b*-PDMS with a molecular weight of 45.5 kg/mol and volume fraction of PDMS 33.5% was used as received. Thin films of PS-*b*-PDMS were obtained by spin-casting toluene solutions of 1.5% by weight of the BCP on the UNCD and silicon nitride substrates, and then the samples were solvent-annealed under toluene vapor at room temperature for 4 h. The annealed film was treated with a 5 s, 50 W CF$_4$ plasma then a 90 W O$_2$ plasma to remove the PS leaving oxygen-plasma-modified PDMS cylinders on the substrate.

PS-*b*-PMMA with a molecular weight of 77 000 g mol^{-1} (M_n=55,000 PS/22,000 PMMA) and a polydispersity of 1.09 was used after removing excess PS homopolymer via Soxhlet extraction in cyclohexane. Thin films were then spin-cast at 5000 rpm from 1.5 wt.-% toluene solution onto clean UNCD and silicon nitride substrates and annealed for six hours at 250° C in an inert atmosphere.

RESULTS and DISCUSSION

Surface roughness of as grown UNCD film is rather high with root mean square (rms) roughness of 7.4 ± 1.3 nm. In a recent paper[24] we have shown that the rms roughness could influence the wetting of BCPs. One simple way to overcome this problem is to use the nucleation side of the UNCD which is nearly atomically smooth with a rms roughness of 0.45 ± 0.035 nm. All the results discussed here are on the nucleation side of UNCD (NU-UNCD).

PS-*b*-PFS

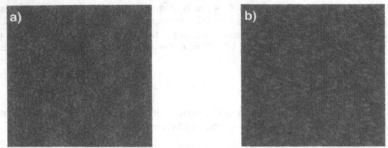

Fig 1. Tapping mode AFM height images of PFS cylinders oriented in-plane on the a) NU-UNCD and b) Si_3N_4 surface. 1% PS-b-PFS was spin cast from Toluene at a spin speed of 4000 rpm for 1 minute and then solvent annealed for 90 mins. Oxygen RIE (7 sccm O_2, 100 W, 20 mT and 3 min) was used to remove the PS matrix prior to imaging. Each image is 2 μm wide.

PS-*b*-PFS is an organic (PS) and organometallic (PFS) BCP. This polymer has been shown to wet a range of substrates, and it has also been successfully used as a lithographic mask for nano- and micron-scale pattern transfer applications. [25][26] On NU-UNCD, a 40 nm film after solvent annealing results in the formation of in-plane cylinders as shown in Fig 1a. Formation of a robust Fe/Si oxide barrier upon exposure to oxygen plasma facilitates the deployment of PFS cylinders shown in Fig. 1 in BCP lithography to develop UNCD nanowires. [24]

PS-*b*-PDMS

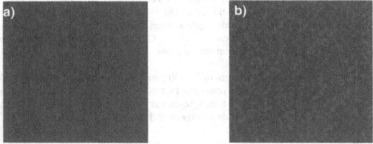

Fig 2. Tapping mode AFM height images of spin cast PS-*b*-PDMS on the a) NU-UNCD and b) Si_3N_4 surface. 1% PS-*b*-PDMS was spin cast from Toluene at a spin speed of 3000 rpm for 1 minute and then solvent annealed for 6 hrs. The annealed film was treated with a 5 s, 50 W CF_4 plasma then a 90 W O_2 plasma to remove the PS matrix prior to imaging. Each image is 2 μm wide.

This BCP is suggested to be an ideal candidate for nanolithography applications as this satisfies two basic criteria i) high value of χ and ii) one of the blocks (PDMS) is highly etch resistant.[27] On a bare NU-UNCD surface this BCP does not wet the substrate well and no significant nanodomains are seen, Fig 2a. On the other hand, using the same film forming and processing conditions on a silicon nitride surface, Fig 2b, in-plane cylindrical structures of PDMS domains are observed. This result indicates that the NU-UNCD surface would require an interlayer such as a thin layer of PS brush to get this polymer to wet the surface.

PS-*b*-PMMA

The self-assembly of this BCP has been well characterized, mostly on Si-based substrates. Control over the ordering of PMMA cylinders has been demonstrated via guided self-assembly such as graphoepitaxy. A thin film of this polymer on silicon nitride forms arrays of in-plane cylinders after thermal annealing, Fig 3b, whereas, at the same processing conditions on NU-UNCD the cylinders seem to be aligned vertically, Fig 3a. It appears that the preferential interaction of one block with the substrate and/or the difference of surface tension between two block components leads to the vertical/parallel orientation of cylindrical microdomains in the vicinity of two interfaces at the substrate/polymer and polymer/air. Further experiments, at various film thicknesses, inclusion of a wetting layer, and varying the annealing conditions may enable control over the orientation of these cylinders with respect to the UNCD surface.

Fig 3. Tapping mode AFM height (a) and phase (b) images of spin cast PS-*b*-PMMA on the a) NU-UNCD and b) Si$_3$N$_4$ surface. 1.5 % PS-*b*-PMMA was spin cast from Toluene at a spin speed of 5000 rpm for 1 minute and then thermal annealed for 6 hrs at 250° C. Each image is 2 μm wide.

CONCLUSIONS

In this work we have surveyed three cylinder-forming BCPs for their self-assembly on NU-UNCD surfaces. All these polymers formed cylinders that oriented in-plane on Si$_3$N$_4$ substrates. At the same experimental conditions only PS-*b*-PFS formed in-plane cylinders on NU-UNCD. PS-*b*-PDMS suffered from wetting related issues on a bare NU-UNCD whereas the PMMA cylinders in PS-*b*-PMMA are oriented vertically. These results indicate that besides PS-*b*-PFS other BCPs would require modifications in addition to deposition on a smooth surface. Thus, to be considered as a candidate for BCP lithography, for instance to develop nanowires, the UNCD might require pre-treatment with an interlayer that would facilitate the wetting, ordering and orientation of BCPs.

ACKNOWLEDGMENTS

Use of the Center for Nanoscale Materials was supported by the U. S. Department of Energy, Office of Science, Office of Basic Energy Sciences, under Contract No. DE-AC02-06CH11357.

13

REFERENCES

1   Y. Cui, Q. W. Wei, H. Park and C. M. Lieber, *Science* **293:**1289-1292 (2001).

2   G. Zheng, F. Patolsky, Y. Cui, W. U. Wang and C. M. Lieber, *Nature biotechnology* **23:**1294-1301 (2005).

3   Y. Ishimori, *Anal Chem* **66:**3830-3833 (1994).

4   T. Strother, W. Cai, X. S. Zhao, R. J. Hamers and L. M. Smith, *J Am Chem Soc* **122:**1205-1209 (2000).

5   F. J. Yusta, M. L. Hitchman and S. H. Shamlian, *J Mater Chem* **7:**1421-1427 (1997).

6   N. Yang, H. Uetsuka, E. Osawa and N. C. E, *Angew Chem Int Ed* **47:**5183-5185 (2008).

7   H. Shiomi, *Jpn J Apply Phys* **36:**7745-7748 (1997).

8   Y. S. Zou, T. Yang, W. J. Zhang, Y. M. Chong, B. He, I. Bello and S. T. Lee, *Apply Phys Lett* **92:**(2008).

9   D. M. Gruen, *MRS Bulletin* **23:**32-35 (1998).

10   O. A. Shenderova and D. M. Gruen, *Ultra Nanocrystalline Diamond - Synthesis, Properties, and Applications*, William Andrew Publishing, New York (2006).

11   P. Mansky, P. M. Chaikin and E. L. Thomas, *J Mater Sci* **30:**1987-1992 (1995).

12   F. Weigla, S. Frickera, H.-G. Boyena, C. Dietricha, B. Koslowskia, A. Plettla, O. Purschea, P. Ziemanna, P. Waltherb, C. Hartmannc, M. Ottc and M. Möller, *Diamond and Related Materials* **15:**1689-1694 (2006).

13   J. P. Spatz, S. Mo¨ssmer, C. Hartmann, M. Mo¨ller, T. Herzog, M. Krieger, H.-G. Boyen, P. Ziemann and B. Kabius, *Langmuir* **16:**407-415 (2000).

14   A. H. Gabor, E. A. Lehner, G. Mao, L. A. Schneggenburger and C. K. Ober, *Chem Mater* **6:**927-934 (1994).

15   D. Sundrani, S. B. Darling and S. J. Sibener, *Langmuir* **20:**5091-5099 (2004).

16   D. Sundrani, S. B. Darling and S. J. Sibener, *Nano Letters* **4:**273-276 (2004).

17   S. O. Kim, H. H. Solak, M. P. Stoykovich, N. J. Ferrier, J. J. D. Pablo and P. F. Nealey, *Nature* **424:**411-414 (2003).

18   C. J. Hawker and T. P. Russell, *MRS bulletin* **30:**952-966 (2005).

19   R. A. Segalman, K. E. Schaefer, G. H. Fredrickson, E. J. Kramer and S. Magonov, *Macromolecules* **36**:4498-4506 (2003).

20   R. A. Segalman, H. Yokoyama and E. J. Kramer, *Adv Mater* **13**:1152-1155 (2001).

21   S. B. Darling, *Progress in Polymer Science* **32**:1152-1204 (2007).

22   A. V. Sumant, O. Auciello, H.-C. Yuan, Z. Ma, R. W. Carpick and D. C. Mancini, Large Area Low Temperature Ultrananocrystalline Diamond (UNCD) Films and Integration with CMOS Devices for Monolithically Integrated Diamond MEMS/NEMS-CMOS Systems, in *Proc of SPIE - Micro- and Nanotechnology Sensors, Systems, and Applications*, ed by M. S. I. Thomas George, Achyut K. Dutta,, pp. 731817-731811 - 731817-731817 (2009).

23   A. V. Sumant, D. S. Grierson, J. E. Gerbi, J. A. Carlisle, O. Auciello and R. W. Carpick, *Phys Rev B* **76**:235439-235442 (2007).

24   M. Ramanathan, S. B. Darling, A. V. Sumant and O. Auciello, *J Vacuum Sci Tech - A* **28**:ASAP (2010).

25   M. Ramanathan, E. Nettleton and S. B. Darling, *Thin Solid Films* **517**:4474-4478 (2009).

26   M. Ramanathan and S. B. Darling, *Soft Matter* **5**:4665-4671 (2009).

27   Y. S. Jung and C. A. Ross, *Nano Lett* **7**:2046-2050 (2007).

Mater. Res. Soc. Symp. Proc. Vol. 1203 © 2010 Materials Research Society   1203-J16-02

Simulations of CVD Diamond Film Growth Using a Simplified Monte Carlo Model

Paul W. May[1], Jeremy N. Harvey[1], Neil L. Allan[1], James C. Richley[1] and Yuri M. Mankelevich[2]

[1]School of Chemistry, University of Bristol, Bristol BS8 1TS, United Kingdom.

[2]Skobel'tsyn Institute of Nuclear Physics, Moscow State University, Vorob'evy gory, Moscow 119991, Russia.

ABSTRACT

A simple 1-dimensional kinetic Monte Carlo (KMC) model has been developed to simulate the chemical vapour deposition (CVD) of a diamond (100) surface. The model considers adsorption, etching/desorption, lattice incorporation, and surface migration along and across the dimer rows. The reaction probabilities for these processes are re-evaluated in detail and their effects upon the predicted growth rates and morphology are described. We find that for standard CVD diamond conditions, etching of carbon species from the growing surface is negligible. Surface migration occurs rapidly, but is mostly limited to CH_2 species oscillating rapidly back and forth between two adjacent radical sites. Despite the average number of migration hops being in the thousands, the average surface diffusion length for a surface species before it either adds to the diamond lattice or is removed back to the gas phase is <2 sites.

INTRODUCTION

Chemical vapour deposition (CVD) of diamond is a maturing technology that is beginning to find many commercial applications in electronics, cutting tools, medical coatings and optics [1]. The CVD process involves the gas phase decomposition of a gas mixture containing a small quantity of a hydrocarbon in excess hydrogen [2]. A typical gas mixture uses CH_4 in H_2 (plus sometimes additional Ar or N_2), and depending upon the growth conditions, substrate properties and growth time, this produces polycrystalline films with grain sizes from ~5 nm to mm. Films with grain sizes less than 10-20 nm are often called ultrananocrystalline diamond films, UNCD; those with grain sizes a few 10s or 100s of nm are nanocrystalline diamond films (NCD); those with grain sizes microns or tens of microns are termed microcrystalline diamond films (MCD) range; and those with grain sizes approaching 1 mm are single crystal diamond (SCD).

However, to obtain a diamond film with the desired morphology combined with controlled electronic and mechanical properties requires a detailed understanding of the many parameters affecting growth, such as the substrate temperature, gas mixture, process pressure, *etc.* The difficulty with this is that, even 20 years after diamond CVD was first developed, the exact details of the growth mechanism remain controversial. The so-called 'standard growth mechanism' [3] developed in the early 1990s is a reasonably robust description of the general CVD diamond process. In this model, atomic H, created by thermal or electron-impact dissociation of H_2, is the driving force behind all the chemistry. It is widely accepted [4,5] that the main growth species in standard diamond CVD is the CH_3 radical, which adds to the diamond surface following hydrogen abstraction by H atoms. An elevated substrate temperature (typically >700°C) allows migration of the adsorbed C species until they meet a step-edge and add to the diamond lattice. Another role for the atomic H is to etch back into the gas phase any adsorbed carbon groups that have deposited as non-diamond phases. It is believed that hydrocarbons C_xH_y with 2 or more carbons ($x \geq 2$) are prevented from contributing to the growth by the 'β-scission' reaction which is a rapid, low energy, efficient process that stops the build up of polymer chains on the growing surface. Diamond growth is therefore seen as competition between etching and deposition, with carbons being added to the diamond on an atom-by-atom basis.

Our group recently developed a modified version of the standard growth model which considers the effects of all the C_1 hydrocarbon radicals (CH_3, CH_2, CH and C atoms) on both monoradical and biradical sites on a (100) diamond surface [6]. Our growth model also relies upon surface migration of CH_2 groups along and across the reconstructed dimer rows in order to predict growth rates to within a factor of two of experimental observations. Using the model we derived expressions for the fraction of surface radical sites based upon the substrate temperature, T_s, and the concentrations of H and H_2 above the surface. Under typical CVD diamond conditions with T_s~900°C and 1%CH_4/H_2 around 10% of the surface carbon atoms support radical sites.

Despite these successes, evidence for surface migration, nucleation processes, the effects of gas impurities and gas-surface reactions are sparse and mostly circumstantial. To investigate these ideas we developed a simplified one-dimensional Monte Carlo (MC) model of the growth of diamond films [7] for a fixed set of process conditions and substrate temperature. Although the model was only 1D, the interplay between adsorption, etching/desorption and addition to the lattice was modelled using known or estimated values for the rates of each process. For typical CVD diamond conditions, the model predicted growth rates of ~1 μm h^{-1}, consistent with experiment. Various other growth processes were also predicted, such as step-edge growth, a large positive value for the Ehrlich-Schwoebel potential for migrating species attempting to migrate off the top of step-edges leading to atomic-scale 'wedding cake' structures, and it also showed that β-scission is not as important for determining the surface morphology as previously envisaged.

In a follow-up paper [8], we modelled surface defects by assigning values for the probability of their appearance following certain surface processes, such as migration and adsorption. Such immobile, unetchable surface defects acted as critical nuclei, allowing the nucleation of new layers, and thus a greatly increased growth rate when the rate-determining step for growth is new layer nucleation. The defects also instigate the (re)nucleation of a new crystallite, ultimately leading to a polycrystalline film. We showed that using these ideas, we could qualitatively model columnar growth of MCD films, as well as NCD and UNCD morphologies.

However, these MC models relied heavily upon the kinetic parameters for the various surface process reported in the literature. To extend the MC model further, for example to include temperature dependence, it was necessary to re-examine these values to determine their accuracy and consistency with the microscopic rates for elementary processes at the diamond surface. In this paper we shall re-examine the processes of CH_3 adsorption, surface migration, and etching and try to rationalise models for their temperature dependent rates which will then be used in a new version of the MC program.

THEORETICAL METHODS

The original model for the MC program is given in detail in refs.[7] and [8] and therefore we shall give only a brief description here, along with new additions and modifications. In our MC model, the (100) diamond lattice is represented in only 2 dimensions, as a cross-section, with the top (growing) surface positioned towards the top of the screen. Each C atom is represented by a square block within the lattice, with different coloured blocks representing different 'types' of carbon bonding. Carbons that are fully bonded into the bulk diamond lattice are coloured dark-blue whereas carbons that form the surface layer are coloured grey. Green blocks are used to represent pendant CH_3 groups or bonded CH_2 structures that bridge along or across the rows of the dimer pairs on the reconstructed (100) surface, and these are immobile. A new modification is that now we allow an immobile green block to become 'activated' following a successful H-abstraction reaction. Such activated blocks are coloured red, and are allowed to migrate to a neighbouring block, so long as there is a surface radical site present. This change has been implemented because computational models of carbon migration [9,10] suggest that dual activation of the migrating and neighbouring sites is required. A surface radical site is coloured magenta, and occurs as a result of a grey block being activated by a successful H abstraction (see figure 1).

Figure 1. A schematic diagram of the model for the cross-section of the diamond surface and some of the processes. Magenta blocks (M) represent activated surface radical sites. The unlabelled light-grey blocks represent unactivated, unreactive (hydrogenated) surface sites, while dark-blue blocks represent bulk (sub-surface) diamond. Green blocks (G) represent immobile CH_3 or CH_2 groups created as a result of adsorption of CH_3 from the gas phase onto M sites (process labelled 1). The red blocks (A and B) represent activated CH_2 groups that are able to migrate. In process 2, red block B can jump left or right since there is an M block at either site. In process 3, red block A cannot jump right since there is no M block there. But it can jump left and drop down the step-edge (following the 'lemmings' scenario [6]) since there is an available M block at the corner.

The grid has a maximum size of 600×400. At the start of the program, a flat horizontal surface of grey blocks is defined at the bottom of the screen to represent the surface of a single crystal diamond substrate. This new version of the program is now fully stochastic (so that the MC program may now be considered a true KMC model), and operates by comparing the relative rates of each process rather than the probabilities of each processes occurring compared to the fastest in previous versions. The program now generates a random number, R ($0 \leq R < 1$), so that at each simulation step a process is chosen with a probability proportional to its rate.. The randomly chosen process is carried out, along with any consequences, and a new list of possible processes is generated ready for the next random number comparison. The processes involved are:

(a) Surface site activation. A grey surface block is activated by H abstraction to form a surface radical site. The grey block then turns magenta, and this square is now available for adsorption of an incoming green or a migrating red block.

(b) Surface site deactivation. This is the opposite to (a), in that a magenta surface radical site is deactivated by H addition to become a standard unreactive grey surface site.

(c) Adsorption of a CH_3 group onto a surface radical site. In this case a new incoming green-coloured block is chosen at a random horizontal position corresponding to one of the activated surface sites (red or magenta) at the top of the screen, and then allowed to drop vertically until it meets the surface, whereupon it temporarily adsorbs at this position. This block represents a generic C_1 adsorbing unit, which is most probably CH_3 but could be species such as C, CH, CH_2 or even CN, as favourable processes for addition of these species to activated sites exist. (In fact, C, CH and $^1CH_2$ can undergo facile processes for addition even to an *unactivated* (grey) site [11] but these processes do not play a major role under the present conditions.) The adsorbed green block then has a number of possible pathways (d)-(h), depending upon the local morphology where it landed, and each possible fate is included in the list of possible processes. One other possible fate for it is to stick permanently to form a static, unetchable defect – however, in the work described here we have turned off this option since we are focusing upon the other processes.

(d) Etching. Isolated CH_2 bridging units or CH_3 groups may be etched back into the gas phase following H abstraction reactions. The green block is then removed and forgotten by the program.

(e) Activation. As a result of a subsequent H abstraction, the CH_3 becomes an activated CH_2 group (and the green block turns red) which is now capable of migration.

(f) Deactivation. As a result of H addition onto an activated CH_2 group, the group is 'deactivated' and returns to being an immobile (green) CH_2 bridge or pendant CH_3.

(g) Migration. An activated (red) CH_2 block may jump sideways left or right one position, so long as there is a (magenta) radical site available to jump into. If migration occurs, the block jumps to the neighbouring site (and remains red), and the site it previously occupied now become magenta, since this is now an activated surface site.

(h) Addition to the lattice. If an adsorbing block lands adjacent to a step edge, it will fuse to the lattice and turn grey [12]. This is an example of an Eley-Rideal-type process (ER). Alternatively, if a migrating red block jumps and lands next to a step-edge, it too may fuse to the lattice and turn grey. This is a Langmuir-Hinshelwood-type process (LH).

(i) Once a block is no longer part of the surface layer, *i.e.* it has been buried beneath at least one other layer it turns dark-blue to represent the bulk lattice.

Three other features of the model need to be mentioned. First, this 1D model assumes that the 'normal critical nucleus' for diamond growth is two adjacent blocks. This is defined as the smallest immobile, unetchable surface feature that provides step-edges suitable for propagating layer growth. Under standard growth conditions a normal 2-block critical nucleus can be formed by (i) an ER-type process, where an incoming green block adsorbs directly next to a previously adsorbed block causing both of them to bond together, or (ii) a LH-type process where an adsorbed red block migrates next to a green or red block and they fuse together. These two processes form the basis for new layer nucleation in the absence of defects.

Second, β-scission is modelled by scanning the surface blocks after every time-step and identifying and deleting any 2-block pillars that may have arisen as a result of blocks landing or migrating.

Finally, there is the issue of blocks migrating off the top of step-edges. Previously, [7,8] we adopted the 'cowards' scenario as the default process, which meant that migrating blocks could not jump off the top of step-edges, consistent with a positive Ehrlich-Schwoebel barrier for this process. However, recent quantum mechanical calculations [13] suggest that this barrier is much smaller than previously thought, and is of a similar magnitude to the barrier for migration on a flat surface. Therefore we have now adopted the 'lemmings' scenario as the default process, whereby migrating (red) blocks can readily jump off the step-edge and 'fall' to the bottom (which may be several blocks in height), landing in the bottom corner (so long as the surface block beneath is activated, i.e. magenta). The block then fuses to the lattice at this corner.

The program was run until it was stopped manually or until a preset number of layers (typically 300 to provide statistical invariance) had grown, at which point the data were saved. Depending upon the input parameters for the various events, the program took several hours to grow 300 layers (on a Pentium 4 PC). At each step the time taken was updated according to $t_{new} = t_{old} - \ln (R)/S$, where t_{old} is the cumulative time up to the previous step, R is a random number $(0 \leq R < 1)$, and S is the sum of the rates of all possible processes [14]. Thus, the growth rate can be calculated since in the simulation 300 layers of diamond grew in this time, with the average C-C distance along a (100) diamond face (i.e. 1 block) being 0.0892 nm.

In this paper the growth conditions were fixed for standard polycrystalline CVD grown using 1%CH_4/H_2 at a process pressure of ~20 Torr but with varying temperatures [6]. We shall now reinvestigate each of the processes in turn.

CH₃ adsorption

In previous papers we have described a model for the gas chemistry occurring within hot filament or microwave plasma CVD reactors [6,15]. This model has been tested against laser spectroscopy and in situ mass spectrometric measurements. For a given set of process conditions we can use this model to determine, with reasonable accuracy, the concentrations of all the major gas phase species at any position within the reactor. Thus, we can extract the concentration of CH_3 just above the growing surface and extrapolate this to determine the rate of CH_3 species striking the surface per second. The number of CH_3 impacts cm^{-2} s^{-1} is given by $[CH_3]_s \times v/4$, where $v = 3757 \times T_{ns}^{0.5}$ (cm s^{-1}) is the mean thermal velocity of CH_3 and T_{ns} is the gas temperature near the substrate surface. However, most of these impacts will be with a hydrogenated surface C, and so the CH_3 will simply bounce off. Only those impacts which strike dangling bonds will be important for growth and need be considered in the model. We shall ignore the effects of co-adsorbed dopant atoms on the adsorption rate, since this is beyond the scope of the present work [16]. Assuming that the gas temperature near the surface, T_{ns}, is approximately the same as the surface temperature, T_s, and that 1 cm^2 of the diamond surface contains ~1.56×10^{15} C atoms, then the rate at which CH_3 species strike the surface site (in s^{-1}) is given by:

$$CH_3 \text{ impact rate} = \{P \times [CH_3]_s \times (3757 \times \sqrt{T_{ns}}) / 4\} / 1.56 \times 10^{15} \qquad (1)$$

where P is the probability of adsorption onto a radical site (*i.e.* the sticking probability). The value of P results from a combination of factors that reduce the reaction probability, such as a geometrical factor (g) due to unfavourable collision orientation and a steric-electronic factor (s), such that:

$$P = g \times s \tag{2}$$

The factor s can be estimated since it is known from electronic spin statistics that, on average, 3 collisions out of 4 will be on the triplet surface and will not lead to reaction at the high temperatures of diamond CVD [17], and that not all of the surface radical sites will be accessible for adsorption (roughly 50%). This leads to an estimated value for s of ~0.15. For the standard hot filament deposition conditions [6] used in this paper $[CH_3]_s = 1.4 \times 10^{13}$ cm^{-3}, $[H]_s = 1.85 \times 10^{14}$ cm^{-3}, $T_{ns} \sim T_s = 1173$ K and $g = 0.5$, giving a per site rate of CH_3 impact of ~20 s^{-1}. This rate is then multiplied by the number of surface radical sites (red + magenta) available at that time-step to obtain the total relative adsorption rate.

Etching

The etching of diamond in atomic H atmospheres is known to be very slow (0.2–0.5 nm h^{-1} [18]), but nevertheless has been proposed as a mechanism by which surface smoothing occurs during growth [19]. For our MC model, we require the etching rate for an isolated surface CH_2 or CH_3 group, which may be higher than that of the bulk lattice. Simple thermal desorption has been discounted as a removal mechanism due to the relatively low substrate temperatures and the high C–C bond energy. Previously, to obtain a value for the etch rate we followed Netto & Frenklach [20] and assumed that the etching step is simply the reverse of the CH_3 addition process. Here, an adsorbed CH_2 group is removed back into the gas phase (catalysed by H) as CH_3, leaving behind a surface dangling bond. Netto & Frenklach calculated two etch rates for the two types of bridging site (termed A3 and A4 in refs.[9] & [21]), which we previously averaged [7] to get a mean etch rate.

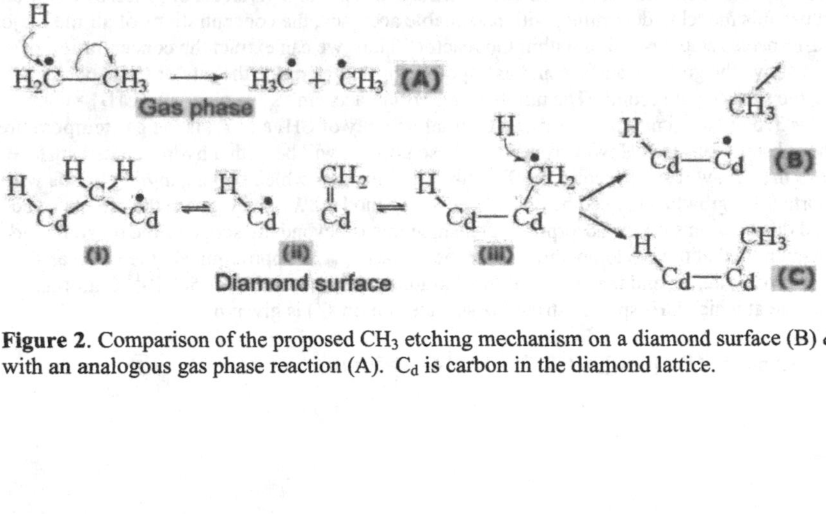

Figure 2. Comparison of the proposed CH_3 etching mechanism on a diamond surface (B) & (C) with an analogous gas phase reaction (A). C_d is carbon in the diamond lattice.

However, the problem with these etch rates is that the assumptions used by Skokov *et al.* [9] in their derivation are questionable. These authors assumed that the H addition reaction to gas phase CH_2CH_3 (figure 2, reaction (A)) is a reasonable analogy to those occurring on the diamond surface (figure 2, reactions (B) & (C)), and thus that the known rate for the former reaction could be used as a good approximation to the etching rate. However, in a gas phase reaction such as this, the excess vibrational energy deposited in the molecule due to formation of a C–H bond can only escape due to relatively slow radiative or collisional processes, with unimolecular decay due to C–C bond cleavage and hence the formation of CH_3 dominating. But on the diamond surface, the heat released by addition of H to the CH_2 group can rapidly be dissipated into the bulk (reaction (C)), so that only prompt C–C bond cleavage (reaction (B)) can compete with vibrational deactivation. The rate of etching (B) will be proportional to the relative lifetime of state (iii), which is small in comparison with the more probable state (i). The rate will also be related to the proportion of the deposited energy which remains close to the surface in state (iii) compared to that dissipated within the timescale required to break the C–C bond. Hence the efficiency of etching by this mechanism can be seen to be inversely dependent on the thermal conductivity of the diamond surface. Diamond has a very high thermal conductivity, and so energy dissipation is likely to be rapid – however, it is not clear if the thermal conductivity in the near-surface region would be as large as that of bulk diamond. Nevertheless, it is likely that the rate of loss of CH_3 would be reduced to such an extent that etching by this mechanism may be essentially negligible (consistent with both the low etch rates [18] and the low values ($<10^{-6}$) of sputtering yield of C atoms per H atoms seen experimentally [22]). Therefore, previous MC models (both ours and others [20,23]) may have significantly over-estimated the etch rates and therefore the importance of etching in controlling surface morphology and growth processes. Work is currently underway to calculate the etch rate using molecular dynamics modelling of the bond breaking and energy dissipation processes, but meanwhile, we have assumed that etching can only occur by direct breaking of the C–C bond. To model this we used an Arrhenius expression for the rate constant for etching

$$k_{etch} = A_{etch} \exp(-E_a / RT_s) \tag{3}$$

where A_{etch} is the collision frequency which we have assumed is the same as that used by Netto & Frenklach (10^{13} s^{-1}), E_a is the activation energy which we have taken to be equivalent to the C–C bond energy (348 kJ mol^{-1}), R is the gas constant and T_s is the substrate temperature. With $T_s = 1173$ K, this gives the per site etching rate as 3×10^{-3} s^{-1}, which is a factor of $1000 \times$ slower than most other processes, confirming the notion that such etching processes are almost negligible.

CH₂ activation and deactivation

From [20], the rate of creation of surface radicals due to of H abstraction is given by

$$\text{Activation rate (s}^{-1}) = k_1 [H]_s U \tag{4}$$

with the rate constant $k_1 = 8.63 \times 10^{-11} \exp(-3360/T_s)$ and U the number of unactivated surface sites (greys + greens).

Also from [20], the rate of deactivating a surface radical site (red block turning green or magenta block turning grey) is

Deactivation rate $(s^{-1}) = k_2 [H]_s A$ (5)

where $k_2 = 3.318 \times 10^{-11}$ cm^3 s^{-1} and A is the number of activated surface sites (reds + magentas).

Surface migration

The migration rate to be considered is that for an activated CH$_2$ bridging group to move along or across a dimer row. Netto & Frenklach [20] obtained a rate constant for these processes to be $\sim 1.5 \times 10^7$ s^{-1} (at $T_s = 900$°C). More recently, Cheesman *et al.* [10] found the activation barrier for hopping to be slightly less than previously thought, with the values for moving along or across the dimer rows being 145.5 and 111.3 kJ mol^{-1}, respectively. Taking an average of these, and assuming the same pre-exponential factor as Netto & Frenklach, we obtain:

$$k_{hop} = 6.13 \times 10^{13} \exp(-128400 / RT_s)$$ (6)

for the rate constant of the pure hopping process. However, the activated CH$_2$ groups will only be able to hop if there is a suitable radical site in a neighbouring position. Previously, to obtain the overall rate of migration (per activated surface CH$_2$ group) we simply multiplied k_{hop} by the chance of a neighbouring site being a radical, typically 0.1. For our standard conditions, this gave values of the rate of migration to be $\sim 1.3 \times 10^7$ s^{-1}, making migration the fastest process by far in the MC model. It also allowed the CH$_2$ group to migrate long distances (10-100 sites) across the surface before being etched or adding to the lattice.

However, there are problems with this simple model, since in reality the rate of migration may be significantly slowed by the lack of availability of surface radical sites. Thus, the migration rate is coupled to the H abstraction rate in a more complex way than we (and others) previously accounted for. The new model takes this into account by only allowing migration to occur if both the CH$_2$ is activated (red) *and* there is a neighbouring activated surface site (magenta) to receive it. One result of this new model for migration is that migrating red blocks hop back and forth rapidly between two adjacent radical sites, and only rarely migrate beyond this when a third surface site activates. Thus, the number of hops made by an individual red block was often of the order of 10^4, but the average surface diffusion length was usually <2 sites.

RESULTS and DISCUSSION

The modified program achieved its goals of simulating the growth of 300 layers of diamond in a few hours, with the morphology continuously evolving on the screen. Figure 3 shows plots of the diamond growth rate and the RMS roughness as a function of T_s for the HFCVD standard conditions, with all other conditions remaining constant. The simulation predicts an increasing growth rate with T_s, as seen in experiment. This is mainly due to an increase in the fraction of surface radical sites, which increases the adsorption rate. The RMS roughness decreases with T_s due to the increased migration of surface species.

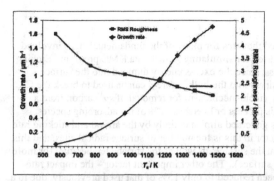

Figure 3. Diamond growth rate and RMS roughness calculated as a function of substrate temperature, T_s, for the standard HFCVD conditions.

The average diffusion length is defined as the mean distance that a migrating species ends up from its initial adsorption site when its migration is permanently terminated by processes such as etching, attachment to the lattice, *etc*. This diffusion length is a function of T_s (see figure 4), mainly through the increase in migration rate. However, the diffusion length remains very small, < 2.5 blocks (equivalent to surface lattice sites) for all temperatures tested. This shows that the major effect of migration is that migrating CH_2 species hop back and forth rapidly between two adjacent radical sites, and only rarely migrate beyond this when a third surface site activates adjacent to one of the previous two. Thus, the number of hops made by an individual red block was of the order of 10^4, but the average surface diffusion distance remained <2 sites.

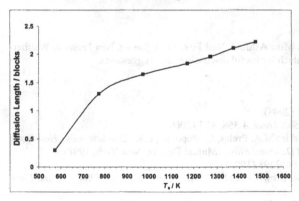

Figure 4. Average diffusion length (in blocks, equivalent to the C–C bond distance on the (100) surface) versus substrate temperature.

CONCLUSIONS

In this paper we have re-evaluated the rates for many of the fundamental steps involved in diamond growth, and which are then used for simulating growth in a KMC program. Etching is now believed to be a negligible process, since the excess energy dumped into the surface groups as a result of H addition can dissipate into the bulk before it can be used to break the C-C bond. This leaves β-scission as the only viable mechanism for removal of sp^3 carbon from a growing diamond surface. However, this process only etches <2% of the adsorbing species, meaning that the diamond growth rate is governed almost entirely by the arrival and sticking rate of carbons onto the surface. A major factor in this is the number of surface radical sites, and this value is governed by the [H]/[H$_2$] ratio at the surface, as well as the surface temperature (or more accurately, the gas temperature near the surface). The other important factor – the impact rate for CH$_3$ species onto the surface – has been reduced to only 15% of that used previously, due to a combination of steric effects and electronic selection rules. This usefully decreases the predicted growth rate to values more in line with those seen in experiment. Migration is now seen as a much more complex process than previously believed, with the surface diffusion length being severely limited by the lack of availability of surface radical sites. Migrating CH$_2$ species can hop back and forth between two adjacent radical sites thousands of times before the migration process is terminated by processes such as the radical sites or CH$_2$ becoming deactivated, the CH$_2$ attaching to a sidewall, *etc*. Thus, the overall average surface diffusion length for a surface species is <2 sites, and this has implications for both the growth rate and the surface roughness.

In future work we shall explore these implications further and investigate the effect of different growth conditions, such as those used to grow SCD or UNCD, upon the predicted growth rates and growth rates and surface morphology.

ACKNOWLEDGMENTS

The authors wish to thank Mike Ashfold, Neil Fox, Keith Rosser, Ben Truscott, Walther Schwarzacher and Michael Frenklach for useful discussions and suggestions.

REFERENCES

1. P.W. May, *Science* **319**, 1490 (2008).
2. P.W. May, *Phil. Trans. Roy. Soc. Lond. A* **358**, 473 (2000).
3. D.G. Goodwin and J.E. Butler, in: M.A. Prelas, G. Popovici, L.K., Bigelow, Eds., *Handbook of Industrial Diamonds and Diamond Films* (Marcel Dekker, New York, 1998).
4. S. J. Harris, *Appl. Phys. Lett.* **56**, 2298 (1990).
5. J.E. Butler, R.L. Woodin, L.M. Brown, P. Fallon, *Phil. Trans. Roy. Soc: Phys. Sci. and Eng.* **342**, 209 (1993).
6. P.W. May, Yu.A. Mankelevich, *J. Phys. Chem. C* **112**, 12432 (2008).
7. P.W. May, N.L. Allan, J.C. Richley, M.N.R. Ashfold, Yu.A. Mankelevich, *J. Phys. Cond. Matter* **21**, 364203 (2009).
8. P.W. May, N.L. Allan, M.N.R. Ashfold, J.C. Richley, Yu.A. Mankelevich, *Diamond Relat. Mater.* (2010), in press (doi: 10.1016/j.diamond.2009.10.030).

9. S. Skokov, B. Weiner, M. Frenklach, *J. Phys. Chem.* **98**, 7073 (1994).
10. A. Cheesman, J.N. Harvey, M.N.R. Ashfold, *J. Phys. Chem. A*, **112**, 11436 (2008).
11. J.C. Richley, J.N. Harvey and M.N.R. Ashfold, *J. Phys. Chem. A* **113**, 11416 (2009).
12. K. Larsson, J.-O. Carlsson, *phys. stat. sol. (a)* **186**, 319 (2001).
13. J.C. Richley, J.N. Harvey, M.N.R. Ashfold, Poster J17.32, *Proc. MRS Fall Meeting* 2009.
14. A.B. Bortz, M.H. Kalos, J.L. Lebowitz, *J. Comp. Phys.* **17**, 10 (1975).
15. Yu.A. Mankelevich, M.N.R. Ashfold, J. Ma, *J. Appl. Phys.* **104**, 113304 (2008).
16. T. Van Regemorter, K. Larsson, *J. Phys. Chem. A* **112**, 5429 (2008).
17. S.J. Klippenstein, Y. Georgievskii, L.B. Harding, *Phys. Chem. Chem. Phys.* **8**, 1133 (2006).
18. R.E. Rawles, S.F. Komarov, R. Gat, W.G. Morris, J.B. Hudson, M.P. D'Evelyn, *Diamond Relat. Mater.* **6**, 791 (1997).
19. R.E. Stallcup II, J.M. Perez, *Phys. Rev. Letts.* **86**, 3368 (2001).
20. A. Netto, M. Frenklach, *Diamond Relat. Mater.* **14**, 1630 (2005).
21. M. Frenklach, S. Skokov, *J. Phys. Chem. B* **101**, 3025 (1997).
22. C.M. Donnelly, R.W. McCullough, J. Geddes, *Diamond Relat. Mater.* **6**, 787 (1997).
23. C.C. Battaile, D.J. Srolovitz, *Annu. Rev. Mater. Res.* **32**, 297 (2002).

Mater. Res. Soc. Symp. Proc. Vol. 1203 © 2010 Materials Research Society   1203-J13-04

Superlattices from diamond

H. Watanabe and S. Shikata

Diamond Research Center, National Institute of Advanced Industrial Science and Technology (AIST), Japan

hideyuki-watanabe@aist.go.jp

Diamond superlattices were fabricated by producing multilayer structures of isotopically pure carbon-12 (^{12}C) and carbon-13 (^{13}C), which confine electrons by a difference in band-gap energy. Secondary ion mass spectrometry (SIMS) measurements were employed to characterize the isotopic composition of the diamond superlattices. Layers between 2 nm and 350 nm in thickness can be designed and fabricated using a microwave plasma-assisted chemical vapor deposition technique.

Introduction

In semiconductor applications, superlattice architectures play an important role in introducing low dimensional properties, and therefore the utilization of novel features such as high electron and hole mobilities, in tuning the emission properties of LEDs and lasers, in the production of optical gratings for dielectric resonator structures, or simply in confining electrons and holes in well defined volume fractions of hetero-junction devices. Superlattice structures (quantum wells) can be engineered by use of different compositions of materials, most prominently, for example, III/V semiconductor junctions such as AlGaAs/GaAs high-electron mobility transistors (HEMTS), where variation of the Al content gives rise to a variation of the electronic band gap[1]. For diamond, an emerging "wide band gap" semiconductor (5.48 eV) superlattice structure has not been realized to date, as diamond cannot be alloyed with other elements. However, as isotopic compositions affect the electronic structure through electron-phonon coupling and through the change of volume with isotopic mass, they can be applied to realize low-dimensional super-lattices from diamond[2]. The excitonic band gap of diamond as a function of mass decreases from ^{13}C to ^{12}C by 19 meV, which

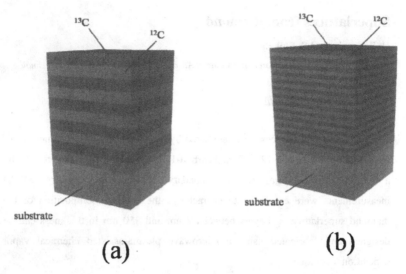

Fig. 1. Schematic sketch of layer sequence of isotopically enriched diamond films. (a) Sample A, had a total of 10 layers. (b) Sample B, had a total of 26 layers.

is more than ten times larger than Si or Ge[3,4,5,6]. In silicon (^{30}Si to ^{28}Si), it is only 1 meV[7], and in germanium (^{76}Ge to ^{70}Ge), it is even less (0.36 meV)[7].

The objective of this study was to find appropriate chemical vapor deposition (CVD) growth conditions for the fabrication of a nanometer-scale ^{12}C/^{13}C multilayered system and to investigate the secondary ion mass spectrometry (SIMS) parameters necessary for depth profiling. Herein, the use of highly purified ^{12}C and ^{13}C diamond superlattices is presented.

Experiment

Two samples of homoepitaxial diamond thin films with a structure consisting of alternating, repeating layers of ^{12}C and ^{13}C diamond of different thickness were investigated, as shown in Figure 1. The samples were grown using a 1.5-kW ASTeX 2.45 GHz microwave plasma-assisted CVD reactor system, enclosed in a stainless-steel vacuum chamber containing a methane (CH_4) and hydrogen (H_2) gas mixture.

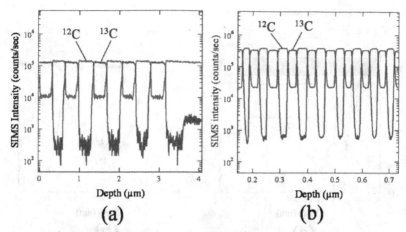

Fig. 2. (a) and (b) SIMS depth profile of the ^{12}C and ^{13}C signals corresponding to the layered structures shown in Figs. 1 (a) and (b), respectively.

Homoepitaxial diamond films were deposited on HPHT synthetic type Ib (001) single-crystal diamond plates (3.0 mm × 3.0 mm × 0.5 mm) with a misorientation angle of less than 0.5° as detected by X-ray diffraction. The chamber pressure was maintained at 25 Torr with a gas mixture 0.15% CH$_4$ in H$_2$, at a total flow rate of 400 sccm. The gas composition was determined from the flow rate of each gas. The applied microwave power was 750 W, and the substrate temperature was kept constant at 800 °C during growth. Isotopically pure methane, either from ^{12}C or ^{13}C, was obtained from Tokyo Gas Chemicals Co. The gases have purities greater than 99.9% for ^{12}C and greater than 98.7% for ^{13}C. Superlattice structures were grown by switching gas lines controlled by mass flow controllers. The layer thickness was controlled by adjusting the deposition time of the growth process, calculated from the growth rate. The growth rate was measured by masking the substrates with diamond plate. After deposition, the step of the grown diamond layer was measured with surface profilometer.

The ^{12}C and ^{13}C depth distribution was analyzed with an ATOMIKA 4500 SIMS Depth Profiling Instrument. The oxygen (O_2^+) primary ion energies (E_p) used were 1 keV and 500 eV at an incident angle normal to the sample surface. For depth profiling,

31

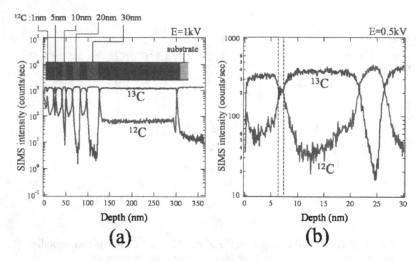

Fig. 3. (a) SIMS depth profile obtained from the test structure of $^{12}C/^{13}C$ multilayer system created on diamond substrate and capped with ^{13}C diamond layer of 5 nm. (b) SIMS depth profile acquired using a 500 eV O_2^+ primary beam.

the primary beam was rastered over an area of 150 μm × 150 μm for $E_p = 1$ keV and 350 μm × 350 μm for $E_p = 500$ eV. The ^{12}C and ^{13}C signals were recorded from the central portion of the raster area only, covering 50% of the area for $E_p = 1$ keV and 37.4% for $E_p = 500$ eV.

Results and Discussion

Figure 2 shows the SIMS depth profiles obtained from the ^{12}C and ^{13}C signals for the 10-layer (sample-A) and 26-layer (sample-B) superlattices shown in Figs.1 (a) and (b), respectively. Depth scales were determined by measuring crater depth using a surface profilometer. As can be seen, the isotopic changes in the crystal form a well-defined pattern which matches the change of gas. The 10 periods of the sample-A superlattice grown on bulk diamond are clearly identifiable in Fig 2(a). The thickness of the ^{12}C and ^{13}C layers derived from the full width at half-maxima (FWHM) values of the SIMS signal is approximately 350 nm. These periodic, synchronous variations in the ^{12}C and ^{13}C composition are also clearly seen in Fig. 2(b), which shows a profile

32

focused nearer to the surface. The variations here occur on the scale of about 30 nm, which is in good agreement with the designed distribution. Thus, Figs. 2(a) and (b) show that both carbon isotope profiles which differ in scale have symmetric properties. This indicates that there is almost no atomic mixing or roughness between the layers. That is, very sharp layers have been formed. These results also suggest that the growth mode of carbon isotopes through the interface for ^{13}C diamond grown on ^{12}C diamond is not essentially different than for ^{12}C diamond grown on ^{13}C diamond.

It is well known that the quantum effect of charge carriers arises from a potential well in the band edges when the well width is comparable to the de Broglie wavelength (λ) of the carriers. The quantum width scale in a semiconductor is given by

$$\lambda = \frac{h}{p}, \tag{1}$$

where $p^2/2m^* = kT$, h is Planck's constant, p is momentum, k is the Boltzmann constant, m^* is electron effective mass, and T is temperature. By taking a value of 0.36 for m^* of an electron, it follows that λ is at least 24 nm for $T = 80$K and 12 nm for $T = 300$K. To obtain the quantum effect from these values of λ, it is necessary to reduce the thickness of a layer to about half the current value.

In order to explore the limits of CVD growth and the SIMS technique, a test structure, a ^{13}C diamond film, containing 5 layers of ^{12}C diamond of varying thickness (1nm, 5nm, 10nm, 20nm and 30 nm) was grown for evaluation using SIMS.

Figure 3(a) shows the SIMS depth profile obtained from the test structure. The ^{12}C composition presence is detectable directly on the surface of the 5 nm ^{13}C diamond cap. This region is highlighted in Fig. 3(b) which was obtained from another area in the test sample using a 500 eV O_2^+ primary beam. Vertical dashed lines indicate the presumed thickness of the ^{12}C layer (1 nm based on deposition time). These results suggest that control of the composition distribution of the 1 nm layer, on the same order as the width of delta-doping, has been achieved with a very high contrast. The results are distributed with a FWHM of about 2 nm located at the center. This pattern corresponds to the resolution of the equipment used.

Conclusions

Superlattice structures were synthesized consisting of a combination of ^{12}C and ^{13}C diamond layers using a microwave plasma-assisted CVD technique. It was found that isotopic diamonds have the possibility to facilitate the band engineering of diamond. The results also indicate the possibility of diamond band engineering with homojunctions, which suggests it may be possible to produce high mobility carriers, which could lead to the development of quantum wells or new potential traps.

Acknowledgments

This work was partially supported by a JSPS Grant-in-Aid for Scientific Research (B) (21360155). The authors thank Christoph E. Nebel (Fraunhofer Institute for Applied Solid State Physics), N. Fujimori (Diamond Research Center, AIST) and K. Kajimura (Technical Research Institute, Japanese Society for the Promotion of the Machine Industry) for their support and encouragement during the early days of the study.

References

[1] H. Ehrenreich, D. Turnbull Eds., Solid State Phys. :Advances in Research and Applications, Vol. 44 (Academic Press, INC. 1991).
[2] H. Watanabe, C. E. Nebel, S. Shikata, Science 324, 1425 (2009).
[3] J. W. Ager III, E. E. Haller, Phys. Stat. Sol. A 203, 3550 (2006).
[4] E. E. Haller, Solid State Commun. 133, 693 (2005).
[5] K. M. Itoh, E. E. Haller, Phys. E 10, 463 (2001).
[6] H. Watanabe, C. E. Nebel, S. Shikata, Proc. New Diamond and Nano Carbons 202 (2008).
[7] M. Cardona, M. L.W. Thewalt, Rev. Mod. Phys. 77, 1173 (2005).

Mater. Res. Soc. Symp. Proc. Vol. 1203 © 2010 Materials Research Society   1203-J05-05

Novel Concepts for Low-Pressure, Low-Temperature Nanodiamond Growth

Using MW Linear Antenna Plasma Sources

Jan Vlček[1&2], František Fendrych[1], Andrew Taylor[1], Irena Kratochvílová[1], Ladislav Fekete[1], Miloš Nesládek[3] and Michael Liehr[4]

[1]Institute of Physics, Academy of Sciences of the Czech Republic, v.v.i, Prague 8, Czech Republic
[2]Institute of Chemical Technology, Department of Physics and Measurements, Prague 6, Czech Republic
[3]IMOMEC division, IMEC, Institute for Materials Research, University Hasselt, Wetenschapspark 1, B3590 Diepenbeek, Belgium
[4]Leybold Optics Dresden GmbH, Dresden, Germany

ABSTRACT

Industrial applications of PE MWCVD diamond grown on large area substrates, 3D shapes, at low substrate temperatures and on standard engineering substrate materials require novel plasma concepts. Based on the pioneering work of the group at AIST in Japan, high-density coaxial delivery type of plasmas have been explored [1]. However, an important challenge is to obtain commercially interesting growth rates at very low substrate temperatures. In the presented work we introduce the concept of novel linear antenna sources, designed at Leybold Optics Dresden, using high-frequency pulsed MW discharge. We present data on high plasma densities in this type of discharge (> 10 E11 cm-3), accompanied by data from OES for CH_4 – CO_2 - H_2 gas chemistry and the basic properties of the nano-crystalline diamond (NCD) films grown.

INTRODUCTION

Typical conditions for PE MWCVD (plasma enhanced microwave chemical vapour deposition) growth of diamond films are a mixture of a hydrocarbon and hydrogen with a very low proportion of hydrocarbon and a substrate temperature of about 600-1000°C [2, 3 & 4]. From the point of view of the material structure and properties, it is important to develop technologies that lead from ultra nanocrystalline diamond – UNCD - (i.e continuously re-nucleating) to nanocrystalline diamond - NCD - (columnar) type of diamond growth [5]. High substrate temperatures restrict the range of suitable substrates and therefore possible applications of these films. Another restriction of PE MWCVD growth is the growth area; typically it is restricted to a maximum diameter of 15cm. Again this not only restricts the range of applications but also their scale. Limitations of HF CVD (hot filament chemical vapour deposition) processes to produce UNCD and NCD films are described in [6]. They relate to the well known difference between thermal HF CVD and PE MWCVD methods. In the first case the reactive species are produced by thermal decomposition of gases at high gas temperatures, PE MWCVD can work far from thermal equilibrium due to reactions induced by energetic particles (ions, electrons and radicals with energies significantly higher than the gas temperature) and therefore creating fluxes of growth species without heating of the gas. This has a fundamental impact on the possibility of low temperature growth. Further more, in resonance cavity PE MWCVD systems used for diamond deposition [7 & 8], due to their high collision rates at high pressures the gas is heated to temperatures of about 3000K. The described system works at pressures of about 1mbar and hence have low collision rates resulting in much lower gas temperatures [1], thus it is suitable for low temperature growth by avoiding the heating of substrates. PE MWCVD plasmas have a further advantage over HF CVD plasmas by avoiding the incorporation of metallic atoms (and inclusions) originating from the filament which are often found in the grown diamond films. Growth rates in HF CVD systems are based on the temperature dependent H-abstraction rates, which lead to extremely slow growth rates at low temperatures [2]. To offset this effect CO_2 can be added into the gas phase [1] allowing a change in the gas chemistry which leads to sufficiently high growth rates at low temperatures. This is not possible with HF CVD systems as CO_2 reacts with the filaments leading to an extremely short lifetime or even etching. Thus our system brings together substantial advantages for low temperature growth on large areas compared to resonance cavity PE MWCVD and HF CVD systems. The aim of this paper is to demonstrate that with the described system high quality NCD films have been produced using a novel pulsed deposition mode and therefore reducing the re-nucleation rate as compared to [1] which used continuous wave plasmas. This is the prime novelty of this work when compared to [1]

EXPERIMENT

For the growth of NCD films a unique Plasma Enhanced Linear Microwave Chemical Vapour Deposition (PELMWCVD) apparatus was built. The PELMWCVD apparatus is capable of producing microwave (2.45GHz) powers to a maximum of 2 x 10kW, with pulse widths from 20 – 100µs. The growth chamber allows the deposition on substrates up to the size of 300mm x 500mm. Microwaves are delivered into the growth chamber in a linear form by four pairs of antennas enclosed in quartz envelopes. The linear microwave plasma sources are arranged parallel to one another above the substrate holder (see Fig. 1).

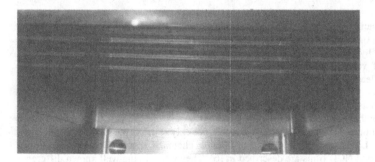

Fig. 1: Microwave plasma sources arranged parallel to one another

Several NCD films have been produced using the following typical process conditions: Gas mixtures – H_2, CH_4 and CO_2 with various ratios, process pressure range – 0.5mBar to 2mBar, microwave power – up to 10.0kW pulsed and growth time of up to 8 hours. Substrates are mounted on a molybdenum holder. During growth, substrate holder temperatures were measured to be in the region of 450°C. Typical substrates used are silicon, quartz and stainless steel. Substrates were seeded with NanoAmando nanodiamond from NanoCarbon Research Institute Ltd using a methodology to reduce clusters and maximize coverage.

OES investigation of the plasma characteristics was carried out using a HORIBA JobinYvon spectrometer (wavelength range of 190nm – 2400nm) with a CCD detector which was connected to the chamber by fiber optics. Focus of the fiber optics was aimed at the central plasma zone above the substrate.

Grown films were investigated with the following techniques: Atomic Force Microscopy (AFM) measurements were performed with a NTEGRA Prima NT MDT system under ambient conditions. Samples were scanned using a HA_NC Etalon tip using semi contact mode. Secondary Electron Microscopy (SEM) images were produced of grown films using a FEI Quanta 3D FEG which combines high resolution scanning electron microscope with focused ion beam (DualBeam). Raman spectroscopy, was carried out at room temperature using a Renishaw InVia Raman Microscope with the following conditions: Wavelength = 488 nm (25 mW), x50 Olympus objective, 65 μm slits, spot focus, Grating = 2400 l/mm. Ellipsometry was carried out using a VASE ellipsometer (Woollam) with a wavelength range of 190nm – 2400nm and an angle of incidence ranging from 20° to 90°.

RESULTS

OES investigation of the plasma revealed that unlike classical systems C2 and CH bands were not present [10]. The concentration of atomic hydrogen was also found to be higher for plasmas with CO_2 addition. Collected spectra from our apparatus showed no lines typical for plasmas during classical CVD diamond growth with pressures of 50 - 100 mbar. Differences between usual emission spectrum and our spectrum are illustrated comparing Fig. 2 [9] and Fig. 3. With CO_2 addition we observed three bands from C-O vibrations with wavelengths at 483.53, 519.82 and 561.02nm (Fig. 4). Hydrogen lines (H_α, H_β, H_γ) and also the hydrogen molecular band around 600nm are visible in the measured spectra.

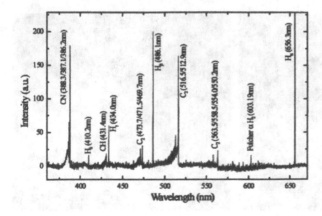

37

Fig. 2: Usual OES spectrum for $CH_4 + H_2$ plasma for classic diamond CVD deposition plasma [9]

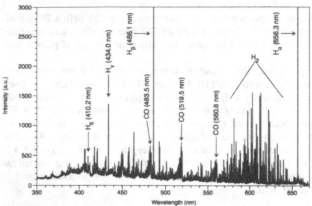

Fig. 3: OES spectrum for $CH_4 + H_2 + CO_2$ plasma from PELMCVD apparatus

AFM showed that for all depositions a continuous layer had been formed in seeded areas. Substrates were purposely prepared with unseeded areas to enable measurement of the layer thickness. It was found that growth rates are enhanced with the addition of CO_2. RMS roughness was measured to be 5-15nm.

SEM investigation was used to evaluate coverage, crystal size, crystal shape and growth rates of all presented NCD layers. Coverage of seeded areas was found to be good for all layers with no holes. The addition of carbon dioxide to process gases resulted in larger crystal size also improved crystalline structure (see: Fig. 4). Cross section images revealed a columnar structure to the layers (see: Fig. 4). Measured growth rates confirm the same trends as in the AFM findings.

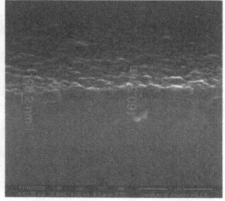

Fig. 4: SEM images of NCD layer with CO_2 addition

Raman spectroscopy showed peaks relating to diamond growth (sp^3) and amorphous carbon (sp^2) were detected for all layers investigated. It was found that the addition of carbon dioxide to process gases significantly improved the intensity and width of the sp^3 Raman peak whilst increasing the sp^3 / sp^2 ratio. (see: Fig. 5). When compared with data reported in [11] it is clear that in this low temperature growth regime sp^3 / sp^2 ratios are greatly improved than for growth with other PE MWCVD and HF CVD systems. A calculation using the Raman factor with relative weights for sp^2 and sp^3 Raman signal gives about 2-3% of sp^2 in our best films which are fully closed and just 50 nm thick.

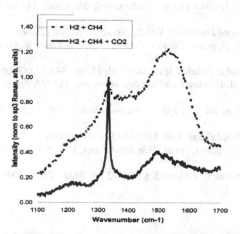

Fig. 5: 488nm Raman spectroscopy of NCD layers showing improvement of sp^3 content and reduction of sp^2 content with the addition of CO_2.

Using the model of RMS / NCD layer / SiO_2 layer / Si layer - ellipsometry confirmed that growth rates were enhanced with the addition of CO_2. Refractive index (n) of layers was found to be 2.40 (400nm) and 2.34 (600nm).

CONCLUSIONS

PELMWCVD apparatus (growth area 30 x 50 cm, 2 x 10 kW max.) was built allowing exploration of non-linear MW absorption plasma chemistry. Suitable low pressure plasma-chemical processes in H_2 + CH_4 + CO_2 gas mixture for enhanced diamond growth rate (20 nm/h) were established for low plasma power densities 3W/cm2 (compared to > 100 W/cm2 for classical MW plasma). Columnar i.e. NCD films with low sp2 (~ 4 % for 100nm), low Rms (5-15nm) and index of refraction (~ 2.4) were prepared.

ACKNOWLEDGMENTS

Thanks to Dr. Masataka Hasegawa from AIST Japan for fruitful discussions. Financial support from the Academy of Sciences of the Czech Republic (grants KAN200100801, KAN300100801, KAN301370701 & KAN400480701) and the European R&D projects (FP7 ITN Grant No. 238201 - MATCON and COST MP0901 - NanoTP) are gratefully acknowledged.

REFERENCES

1. K. Tsugawa, M. Ishihara, J. Kim, M. Hasegawa and Y. Koga: *New Diamond and Frontier Carbon Technology,16, No.6 (2006)*
2. James E. Butler, Anirudha V. Sumant, *Chemical Vapor Deposition, Vol: 14, pp: 145 - 160 (2008)*
3. Nesladek M, Tromson D, Mer C, et al. *Applied Physics Letters, Vol: 88, Issue: 23, Article No: 232111 (2006)*
4. Daenen M, Zhang L, Erni R, et al. *Advanced Materials, Vol: 21, Issue: 6 pp: 670 (2009)*
5. Mares JJ, Hubik P, Kristofik J, et al, *Applied Physics Letters, Vol: 88, Issue: 9, Article No: 092107 (2006)*
6. May P, Smith J, Mankelevich Y. *Diamond & Related Materials, Vol: 15, pp: 345-352 (2006)*
7. Williams O, Nesladek M, Daenen M et al. *Diamond and Related Materials, Vol: 17, Issues 7-10, pp 1080-1088 (2008)*
8. Butler J, Mankelevich Y, Cheesman A et all. *J. Phys.: Condens. Matter, Vol: 21 Art No: 364201 (2009)*
9. T. Vandevelde, T.D. Wu, C. et al. *Thin Solid Films, Vol: 340 pp: 159 - 163 (1999)*
10. Vandevelde T, Nesladek M, Quaeyhaegens C, et al, *Thin Solid Films, Vol: 308, pp: 154 – 158 (1997)*
11. Fortunato W, Chiquito A, Galzerani J et al. *J Mater Sci, Vol: 42 pp: 7331 - 7336 (2007)*

Mater. Res. Soc. Symp. Proc. Vol. 1203 © 2010 Materials Research Society   1203-J05-03

Characterization of nano-crystalline diamond films grown by continuous DC bias during plasma enhanced chemical vapor deposition

V. Mortet[1,2], L. Zhang[3], M. Eckert[4], A. Soltani[5], J. D'Haen[1,2], O. Douhéret[6], M. Moreau[7], S. Osswald[8], E. Neyts[4], D. Troadec[5], P. Wagner[1,2], A. Bogaerts[4], G. Van Tendeloo[3], K. Haenen[1,2].

[1] Institute for Materials Research (IMO), Hasselt University, Diepenbeek, Belgium.
[2] Division IMOMEC, IMEC vzw, Diepenbeek, Belgium.
[3] Electron Microscopy for Materials Science (EMAT), University of Antwerp, Antwerpen, Belgium.
[4] Research group PLASMANT, Department of Chemistry, University of Antwerp, Antwerp, Belgium.
[5] Institut d'Electronique de Microélectronique et de Nanotechnologie, Villeneuve d'Ascq, France.
[6] Service de la Chimie des Materiaux Nouveaux, MateriaNova Research Center, Mons, Belgium.
[7] Laboratoire de Spectrochimie Infrarouge et Raman, Villeneuve d'Ascq, France.
[8] Department of Materials Science and Engineering and A. J. Drexel Nanotechnology Institute, Drexel University, Philadelphia, USA.

ABSTRACT

Nanocrystalline diamond films have generated much interested due to their diamond-like properties and low surface roughness. Several techniques have been used to obtain a high re-nucleation rate, such as hydrogen poor or high methane concentration plasmas. In this work, the properties of nano-diamond films grown on silicon substrates using a continuous DC bias voltage during the complete duration of growth are studied. Subsequently, the layers were characterised by several morphological, structural and optical techniques. Besides a thorough investigation of the surface structure, using SEM and AFM, special attention was paid to the bulk structure of the films. The application of FTIR, XRD, multi wavelength Raman spectroscopy, TEM and EELS yielded a detailed insight in important properties such as the amount of crystallinity, the hydrogen content and grain size. Although these films are smooth, they are under a considerable compressive stress. FTIR spectroscopy points to a high hydrogen content in the films, while Raman and EELS indicate a high concentration of sp^2 carbon. TEM and EELS show that these films consist of diamond nano-grains mixed with an amorphous sp^2 bonded carbon, these results are consistent with the XRD and UV Raman spectroscopy data.

INTRODUCTION

In all times, its excellent properties have always stirred up the interest for diamond. Due to the excellent mechanical properties of this material, smooth nano-diamond thin films could be applied in micro-electro-mechanical systems (MEMS) [1]. Moreover, nano-diamond particles have generated an intense interest in microbiology [2,3] as, for instance, active biocompatible bio-markers. While nano-diamond thin films, i.e. films made of diamond nano particles embedded in an amorphous and non diamond carbon phase, are produced by chemical vapor deposition (CVD) techniques, the most used technique to produce nano-diamond particles is based on a detonation process, which is a cumbersome and hard to control industrial process [2].

On the other hand, the use of substrate bias has been proven to be an efficient method for diamond nucleation and it has been extensively studied. Although nano-diamond thin films grown with substrate bias have been reported in the past [4,5,6]. The recent renewal of interest in nano-diamond and the development of characterization methods urges for new and more elaborate studies.

EXPERIMENTAL

Films were grown on 10 mm in diameter silicon substrates by microwave (MW) plasma enhance (PE) CVD in a NIRIM type reactor [7] during one hour using a 1% mixture of methane diluted in hydrogen. The 15 mm in diameter substrate holder is negatively biased at 260 V. A 40 mm in diameter, grounded stainless steel tube around the substrate holder serves as a counter electrode. Prior to the growth, the sample is first heated to 700°C in a pure hydrogen plasma. Second, the sample is bias enhanced nucleated at a pressure of 50 mbar, a power of 275 W and a negative bias voltage of 260 V using a 1% methane to hydrogen mixture for 10 minutes. The film is then grown under the same bias at a pressure of 80 mbar and a power of 425 W. The samples were characterized by scanning electron microscopy (SEM), atomic force microscopy (AFM), X-ray diffraction (XRD), transmission electron microscopy (TEM), electron energy loss spectroscopy (EELS), and Raman spectroscopy using multiple excitation wavelengths, from infrared (785 nm) to ultraviolet (266 nm).

RESULTS

The mass deposition rate of the films, which show a black and glassy surface (see inset Fig. 1), is high (2 mg.cm^{-2}.h^{-1} – 5.2 μm.h^{-1}), that is nearly 8 times higher than the obtained deposition rate without the application of bias. The large curvature of the sample suggests a high compressive stress. Assuming that there is no plastic deformation of the substrates, a compressive stress higher than 20 GPa has been calculated using the Stoney's equation. These results are similar to the ones reported by Sharda et al. [8].

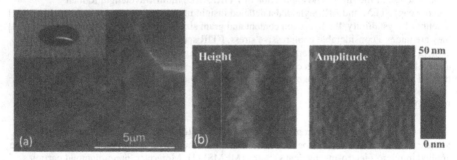

Figure 1. (a) Scanning electron microscopy image of a film grown under a negative DC bias of 260 V– inset: picture of the film. (b) Atomic force microscope images, obtained in tapping mode, of the film on a 500 nm x 500 nm surface.

The surface morphology of the film observed by SEM and AFM is shown in Fig. 1. The surface is smooth (fig. 1a) and a fine structure can be observed by AFM (fig. 1b). The surface roughness is $R_{ms} \sim 4\text{-}5$ nm on a 500 nm x 500 nm surface area, while the observed grain size, which is a convolution of the real grain size and the size of the probing tip, is estimated to be around 20 to 40 nm.

The X-ray diffraction diagram (Fig. 2) of the film only exhibits the diffraction peaks of diamond and the silicon (100) substrate. The diamond peaks are wide and low in intensity. Using the Scherrer's equation with Gaussian and Lorentzian fittings of the peaks, the coherence length of the diamond crystals has been estimated to be 5-6 nm. The diffraction diagram of a 4 nm grain sized powder is also shown on Fig. 2 for comparison.

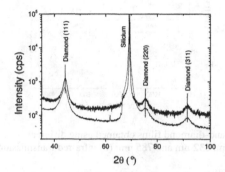

Figure 2. X-ray diffraction diagrams of nano-diamond thin film grown under a negative DC bias of 260 V (upper / black curve) and nano-crystalline diamond powder with a grain size of 4 nm (lower / red curve).

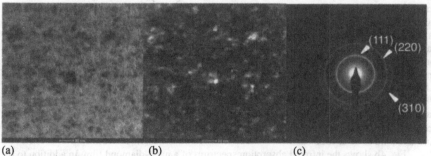

(a)   (b)   (c)

Figure 3. (a) bright field and (b) conical dark field, using the (111) diffraction ring, transmission electron microscopy images and (c) electron diffraction pattern of a nano-diamond thin film grown under a negative DC bias.

The TEM and the electron diffraction pictures (see Fig. 3) support the results obtained by XRD. The dark field image clearly shows that the films consist of nanometer size (~10 nm in average) diamond grains. The sp^3 to amorphous carbon ratio was computed from the EELS spectrum (not shown) and it was determined to be ~ 81%. Raman scattering spectra obtained with different wavelength excitations in the first order range are shown in Fig. 4a. Infra-red and visible Raman scattering spectra exhibit only the graphite (G) band, the disordered graphitic (D) band, and a shoulder at ~ 1150 cm^{-1}. Despite polyacetylene is known to decompose around 400°C, the band at 1150 cm^{-1} is generally assigned to trans-polyacetylene [9-11].

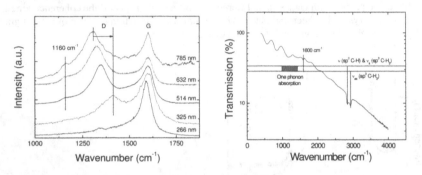

Figure 4. (a) Raman scattering spectra of the nano-diamond films obtained using different excitation wavelengths; 266 nm, 325nm, 514 nm, 632 nm and 785 nm; (b) infra-red transmission of a nano-diamond film in a semi-log plot (b).

The D band, that is due to double Raman resonant scattering [12,13], decreases in relative intensity vs the G band while shifting to higher wavenumbers at a rate of ~ 60 cm^{-1}/eV as the photon energy (E_{ph}) of the excitation laser increases. The D band, that should be located at 1495 cm^{-1}, is not significantly observed in UV Raman scattering at E_{ph} = 4.66 eV (266 nm). Nonetheless, the tail at lower wavenumbers of the G band might be due to the D band. The UV Raman spectrum shows, besides the G band, a minute and relatively broad band at 1341 cm^{-1}. Raman scattering of nanometer sized diamond grains has already been reported [14, 15]. The diamond band shifts to lower wavelength and broadens as the grain size decreases due to phonon confinement [16]. Yoshikawa et al. [14] have reported a red shift of 13 cm^{-1} and a width of 38 cm^{-1} on detonation diamond with a grain size of 5 nm. The diamond band is also known to shift to higher wavelengths when the material is under compressive stress at a rate of 2.9 cm^{-1}/GPa [17]. Thus, the tiny band at 1341 cm^{-1} is not assigned to the D band but to nano-diamond under compressive stress. The diamond grain size has been estimated to be 7-8 nm and a compressive stress of σ ~ 6.5 GPa is calculated, assuming the peak position of non-stressed nano-diamond to be at 1322 cm^{-1}.

Fig. 4b shows the infrared absorption spectrum of a nano-diamond film. In addition to the thickness interference fringes at low wavelengths (v < 2000 cm^{-1}) and the increasing absorption with wavenumbers due to the presence of graphitic carbon, three significant absorptions bands are observed: between 950 and 1400 cm^{-1} (band I) , at 1600cm^{-1} (band II) and between 2800 and 3000 cm^{-1} (band III). The band III is a convolution of different absorption bands of sp^3 CH$_x$ bonds' stretching vibration modes: the symmetric v_{as}(sp^3 CH$_2$) ~2850 cm^{-1} and antisymmetric v_{as}(sp^3 CH$_2$) ~ 2920 cm^{-1} modes of CH$_2$ [18] and the CH vibration at v_{as}(sp^3 CH) ~ 2830 cm^{-1}

present on the (111) surface of diamond [19]. Using a thin film of polyethylene for the calibration of the CH_2 antisymmetric vibration mode absorption, the atomic concentration of hydrogen in the film is assumed to be higher than 3×10^{22} cm^{-3}. This level of hydrogen is in accordance with other reports on hydrogen concentration in nano-crystalline diamond films [5,20]. High concentrations of hydrogen are generally observed in nano-diamond. It is assumed to be trapped at grain boundaries, thus its concentration increases as the grain size decreases [20]. Classical Molecular Dynamics modeling shows a large increase of dangling bonds formation at high bias voltages that is assumed to enhance the reactivity of the material [20]. It is believed that these dangling bonds contribute to the large deposition rate and the high concentration of hydrogen in the films. The band II is ascribed to either conjugated C=C bonds in polyene, or aromatics rings [21]. The presence of C=C bonds in nano-diamond thin films is in accordance with the HR-EELS results of Michaelson and Hoffman [22]. The band I is attributed to both the one phonon absorption of diamond and the absorption of the deformation (δ) vibration mode of sp^3 CH_x groups. The presence of this CH_x deformation's absorption band is consistent with the high concentration of hydrogen in the film determined through the absorption of CH_2 stretching vibration modes. Furthermore, the CH bending vibration mode has been also observed by HR-EELS on nano-diamond films [22]. Although the one phonon absorption is normally forbidden in diamond, the rule breaks in case of diamond grains of very small size [23]. One can also notice that absorption bands of trans-polyacetylene at ~1000 cm^{-1}, 1800 cm^{-1} and ~3045 cm^{-1} are not observed on the IR transmission spectrum, and the Raman band at 1160 cm^{-1} does not significantly shift with the excitation energy as expected for trans-polyacetylene [24]. Nonetheless, this band has to be assigned to a CH vibration mode as it has been shown by deuteration experiments [22]. Thus, it is strongly believed that this band has to be assigned to a deformation mode of CH_x bonds only.

CONCLUSIONS

Structural properties of diamond thin films grown by continuously bias assisted MW PE CVD have been characterized using different techniques. All characterization methods confirm that nano-crystalline diamond films have been grown. The films consist of nano-diamond grains of ~ 10 nm in size mixed with amorphous sp^2 bonded carbon. They are under large compressive stress and present a high concentration of hydrogen. Visible and IR Raman scattering results show only the signature of sp^2 bonded carbon while the diamond signature is only observed using UV excitation scattering. Furthermore, these results disagree with previous reports, which conclude that the Raman scattering band at ~1150 cm^{-1} is due to trans-polyacetylene. In contrast, it is proposed that the origin of this band can be considered to be a deformation mode of CH_x bonds

ACKNOWLEDGMENTS

This work was financially supported by the Research Programs G.0068.07 and G.0430.07 of the Research Foundation - Flanders (FWO), the Methusalem "NANO network Antwerp-Hasselt", and the IAP-P6/42 project 'Quantum Effects in Clusters and Nanowires'.

REFERENCES

1. O.A. Williams, M. Nesládek, M. Daenen, Sh. Michaelson, A. Hoffman, E. Ōsawa, K. Haenen, R.B. Jackman, Diamond Relat. Mater. **17**, 1080 (2008).
2. A. Krueger, Advanced Mater. **20**, 2444 (2008).
3. H. Huang, Pierstorff, E. Osawa, D. Ho, Nano Lett. **7**, 3305 (2007).
4. N. Jiang, K. Sugimoto, K. Eguchi, T. Inaoka, Y. Shintani, H. Makita, A. Hatta, A. Hiraki, J. Cryst. Growth **222**, 591 (2001).
5. M. Schreck, T. Baur, R. Fehling, M. Muller, B. Stritzker, A. Bergmaier, G. Dollinger, Diamond Relat. Mater. **7**, 293 (1998).
6. C.Z. Gu, X. Jiang, J. Appl. Phys. **88**, 1788 (2000).
7. V. Mortet, M. Daenen, T. Teraji, A. Lazea, V. Vorlicek, J. D'Haen, K. Haenen, M. D'Olieslaeger, Diamond Relat. Mater. **17**, 1330 (2008).
8. T. Sharda, T. Soga, T. Jimbo, M. Umeno, Diamond Relat. Mater. **10**, 352 (2001).
9. A.C. Ferrari, J. Robertson, Phys. Rev. B **63** (2001) 121405(R).
10. H. Kuzmany, R. Pfeiffer, N. Salk, B. Gunther, Carbon **42** (2004) 911.
11. T. Lopez-Rios, E. Sandre, S. Leclercq, E. Sauvain, Phys. Rev. Lett. **76** (1996) 4935.
12. S. Reich, C. Thomsen, Phil. Trans. R. Soc. Lond. A **362**, 2271 (2004).
13. C. Thomsen, S. Reich, Phys. Rev. Lett. **85**, 5214 (2000).
14. M. Yoshikawa, Y. Yuri, M. Maegawa, G. Katagiri, H Ishida, A. Ishitani, Appl. Phys. Lett. **62**, 3114 (1993).
15. J. Chen, S.Z. Deng, J. Chen, Z.X. Yu, and N.S. Xu, Appl. Phys. Lett. **74**, 3651 (1999).
16. R.J. Nemanich, S.A. Solin, R. M. Martin, Phys. Rev. B **23**, 6348 (1981).
17. H. Boppart, J. van Straaten, I. Silver, Phys. Rev. B **32**, 1423 (1985).
18. G. Socrates, Infrared and Raman characteristic group frequencies – tables and charts, 3rd ed. (John Weiley & Sons Ltd, Chichester, 2001) p. 50.
19. B.F. Mantel, M. Stammler, J. Ristein, L. Ley, Diamond Relat. Mater. **9**, 1032 (2000).
20. Sh. Michaelson, O. Ternyak, R. Akhvlediani, O.A. Williams, D. Gruen, A. Hoffman, Phys. Stat. Sol. (a) **204**, 2860 (2007).
21. Bernhard Schrader, Infrared and Raman spectroscopy – Methods and applications (VCH Verlagsgesellschaft mbH, Weinheim, 1995), p.197-199.
22. Sh. Michaelson and A. Hoffman, Diamond Relat. Mater. **15**, 486 (2006).
23. P. K. Bachmann and D. U. Wiechert, Diam. Relat. Mater. **1**, 422 (1992).
24. J.-Y. Kim, E.-R. Kim, D.-W. Ihm, M. Tasumi, Bull. Korean Chem. Soc. **23**, 1404 (2002).

Mater. Res. Soc. Symp. Proc. Vol. 1203 © 2010 Materials Research Society   1203-J16-01

Effect of Substitutional or Chemisorbed Nitrogen on the Diamond (100) Growth Process

Karin Larsson

Department of Materials Chemistry, Uppsala University, Box 538, 751 21 Uppsala, Sweden

ABSTRACT

The present paper outlines the energetic and kinetic effect by substitutional N, or by coadsorbed NH_x (x =1, 2), on one of the key growth steps in the CVD growth mechanism of diamond (100); H abstraction by gaseous H radical species from the (100) surface plane. Theoretical calculations were performed based on Density Functional Theory under periodic boundary conditions. Substitutionally positioned N was shown to have a large effect on the H abstraction process. The H abstraction energy from the diamond surface was greatly improved with N positioned in C layer 2. In order to outline the effect by N on the growth rate, the barriers of energies were calculated. The barrier of abstraction was shown to substantially decrease with N substitutionally positioned in the second C layer, leading to an improvement of the abstraction reaction rate by approximately a factor of 3.

INTRODUCTION

The growth of diamond films with desired properties and morphology requires a perfect recognition of the parameters affecting the growth process [1-3]. It is especially crucial to understand how these parameters will affect the growth on an atomic level. In the early 1990s, Locher et al. reported, for the first time, major changes in the polycrystalline diamond CVD growth by introducing a small percentage of nitrogen in the gas phase [4]. In a more recent publication, Achard et al. performed a careful study on the effects induced by the presence of N during the (100) single crystal diamond growth process in order to understand how it affects the growth rate and the surface morphology [5]. In accordance with their observations, they proposed that sub-surface nitrogen, through an electron transfer to a surface C radical, is probably responsible for the important effect observed experimentally. This assumption was initially made by Frauenheim [6] et al. who showed theoretically that this specific electron transfer largely affects the reaction energies and barriers of the different steps involved in the growth mechanism proposed by Harris and Goodwin for diamond(100) [7-8]. This mechanism is composed of 5 different steps: a) surface H abstraction by a gaseous H radical; b) CH_3 adsorption on the newly formed surface radical; c) H abstraction from CH_3 by a gaseous H radical; d) C-C dimer opening by a b-scission rearrangement; and e) the final formation of a six carbon ring. Independent of the surface reaction model considered, the importance of i) H abstraction from the diamond surface, ii) CH_3 adsorption to the newly formed surface radical site, and iii) a following H abstraction, forming an adsorbed CH_2, has earlier been stressed [9]. It is therefore of the greatest interest to analyze the effect of substitutional nitrogen on these reaction types, generally considered as initial CVD growth steps of diamond in an H /CH_3 atmosphere.

The present author has recently published a series of theoretical papers were the effect of substitutional N on the adsorption, surface migration and incorporation of C precursors (in the form CH_3 or CH_2) into diamond (100) [10-12]. These earlier studies were predominantly thermodynamic studies at zero Kelvin, and the main focus was to study the structural evolution and possibilities for certain reactions to take place (i.e. influence of N on chemical driving forces). However, the initial diamond growth processes are very much dominated by H abstraction reactions from the diamond (100) terrace. The purpose with the present study is to, by using density functional theory (DFT) under periodic boundary condition, achieve a deeper *energetic and kinetic* understanding of the N effect on the H abstraction processes by considering different nitrogen positions on, or within, the diamond (100) surface.

METHODS

Density Functional Theory (DFT), using the program package CASTEP from Accelrys, Inc., was used for the energy and geometry optimization calculations [13-14]. The Generalized Gradient Approximation (GGA-PW91) developed by Perdew and Wang, was used in describing the electronic exchange and correlation interactions [15]. GGA-PW91 introduces inhomogeneity by using an electron density gradient expansion. This method is in this sense more accurate than the local density approximation (LDA) method [16-18]. As a result, the GGA method gives more correct total energies and structural energy differences. In the present investigation, the kinetic energy cut off was set to 240 eV in the plane wave function description. A Monkhorst-Pack generated $2 \times 3 \times 2$ k-point mesh was used for all calculations (yielding 6 k-points). This scheme produces a uniform mesh of k-points in the reciprocal space, which has been found to be more reliable than linear or quadratic methods [19]. The geometry optimisation procedure was based on the BFGS algorithm (Broyden-Fletcher-Goldfarb-Sharmo) [20]. The convergence criteria within the calculations were $2 \cdot 10^{-5}$ Ha for the maximum energy change per atom, $4 \cdot 10^{-3}$ Ha/Å for the maximum force per atom, and $5 \cdot 10^{-3}$ Å for the maximum displacement per atom. In addition, the covalent bond strengths were estimated by calculating the electron bond populations by projecting the plane wave states onto the localized basis set by means of Mulliken analysis [21]. The deformation density, i.e., the total density with the density of the isolated atoms subtracted, was also calculated. Regions of positive deformation density indicate the formation of bonds, while negative regions indicate electron loss.

The calculations were based on periodic boundary conditions, where the interfacial models were constructed as periodically repeated super cells. The diamond substrate within the interfacial structures were in the present investigation modelled as a (2×1) reconstructed diamond surface comprised of a 5 layer slab of 12 atoms per layer. The bottom C layer in diamond was hydrogen-terminated with the purpose to saturate the dangling bonds and to maintain the sp^3-hybridization of the carbon atoms. This hydrogenated carbon layer was held fixed to simulate the crystalline bulk structure in the calculations. The other atoms were allowed to fully relax. As a result of earlier test-calculations, the here presented theoretical and model parameters were found adequate to use for the present calculations [22].

RESULTS and DISCUSSION

General

The diamond CVD growth process occurs through a complex set of elementary reaction steps on the surface [22]. The reaction between a gaseous hydrogen radical (H) and the diamond surface is one of the most important step (the H species is crucial for the abstraction of an adsorbed H in order to form a surface carbon radical). The newly formed radical surface site will most probably migrate fast on the surface via migrating H adsorbates, or by the transfer of H from chemisorbed CH_x species. The formation of these surface site radicals are absolutely necessary for the adsorption of gaseous growth species to take place.

When considering the total growth mechanism, it is most probable that the gas phase H abstraction will initially dominate as an elementary reaction step. In order to outline the underlying causes to the experimental observations made (concerning growth rate enhancement when including nitrogen in the CVD reactor chamber), it is therefore of an outermost importance to theoretically investigate the effect of N dopants on this H abstraction process, both energetically and kinetically.

Gaseous H abstraction from diamond terraces.

Several different positions of N have been used in investigating the effect of substitutional N on the H abstraction process. As can be seen in Fig. 1, it is obvious that there is only a minor effect when substituting a first atomic layer C atom with N. However, when replacing C with N in layer 2 and 3, there will be a major effect on the H abstraction energy for both diamond (111) and (100). For diamond (100), the H abstraction energy changed from about -20 kJ/mol (for the non-doped scenario) to a very exothermic value of -140 kJ/mol (with N substitionally positioned in C layer 2). As an explanation to this observation, the resulting diamond (100) surface with a radical C surface site will become substantially stabilized through electronic interactions between N and the radical surface C. In addition, the influence of a co-adsorbed NH_x(x=1 or 2) on the H abstraction reaction, has also been performed. However, the calculations gave here evidence of much lower improvements (at most a decrease by about 80 kJ/mol).

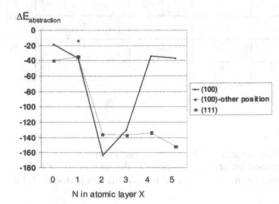

Figure 1. Exothermic H abstraction energies from diamond (100) and (111) surfaces. The numerical values for the non-doped situation are shown as values for the atomic layer 0. Two different, but neighbouring, N positions has been used for the atomic layer 1 within diamond (111).

Energy barriers for gaseous H abstraction from diamond (100).

The reaction energies, as presented in Refs. 10-12, show that N incorporated in the 2nd layer of diamond (100) will improve the driving force for some of the most important growth reactions. In addition, it has in this investigation been found that N will have a large energetic effect on the H abstraction from the surface. However, in order to outline the effect by N on the growth rate, one has also to consider the barriers of energies (i.e. kinetic considerations). One of the main objectives with the present work was hence to study the barrier of abstraction for one of the most important diamond growth steps; H abstraction. As will be presented here, the barrier of H abstraction from the H-terminated diamond (100) surface was shown to substantially decrease with N substitutionally positioned in the second C layer, leading to an improvement of the abstraction reaction rate by approximately a factor of 3.

In calculating the barrier of abstraction for H abstraction, gaseous H was sequentially positioned closer to one of the adsorbed H atoms on the surface (with a direction perpendicular to the surface). This incoming H atom was passing a small energy barrier (TS1) before it was finally adsorbed (weakly) to the surface H. This newly formed H$_2$ "adsorbate" was thereafter allowed to desorb from the surface, leaving a non-terminated surface site. A small energy barrier (TS2) was also here observed. This procedure was repeated for the situations with i) non-doped diamond, ii) N in the 2nd layer, and with iii) NH or NH$_2$ iv) co-adsorbed on the surface. With one exception, all situations showed almost identical energy barriers (both TS1 and TS2). The exception is for the incoming H towards a H-terminated surface with N incorporated in the second C layer. For this situation, no barrier was observed. A summary of the abstraction energies and barriers can be seen in Table I.

Table I. A summary of abstraction energies and barriers (TS1 for the approaching H and TS2 for the outgoing H$_2$) for the different N doping scenarios. An estimation of the rate constants, related to the non-doped situation, is also presented.

	ΔE(abstraction)	ΔE(TS1)	ΔE(TS2))	k / k(no N)
No N	-19	8	8	1
N in layer 2	-163	0	4	~3
NH adsorbate	-29	8	5	~1
NH$_2$ adsorbate	-1	8		~1

CONCLUSION

The present paper presents a theoretical investigation of the energetic and kinetic effect of N (substitutional positioned within the diamond lattice, or co-adsorbed in the form of NH or NH$_2$) on one of the most important key growth steps in the CVD growth mechanism of diamond (100);

gaseous H abstraction from the H-terminated diamond (100) surface. The results, as obtained from calculations based on Density Functional Theory under periodic boundary conditions, can be summarized as follows:

- A radical N in C layer 2 (or 3) will cause improved H abstraction energies. The main reason to this observation is the stabilization of the resulting radical surface C site, which in turn is caused by N-radical C electronic interactions.

- The improvement in H abstraction energy induced by substitutionally positioned N in C layer 2, will lead to an increase in abstraction reaction rate by a factor of approximately 3! This calculated increase in elementary reaction rate is strongly supporting experimental results regarding the effect of N doping on diamond growth rate..

- Radical, or non-radical, forms of N co-adsorbed on the surface does not seem to have any effect on the H abstraction process.

- A very low concentration of N within the surface atomic layers will most probably induce a long range effect, and the reason is twofold. Firstly, surface H atoms have been found to be very mobile (either as chemisorbed species on the surface, or as ligands in e.g. chemisorbed CH_3 species). Secondly, it has here been observed that the adsorption of a gaseous H species to the newly formed radical C surface site (with N in the second C layer) will bind rather strongly to the surface (with a binding energy of -273 kJ/mol)! Hence, there is a very large probability that a newly formed surface radical C site will initiate a chain of H translations on the surface, ending up in a radical C site further away from the initial one. The gain by using N dopants will thereby again be twofold; i) the N dopant will act catalytically since it will never be "consumed", and the adsorption of a CH_3 species to a surface radical site away from the N dopant will not have the possibility to experience the negative effect of the N dopant (see Ref. 23).

ACKNOWLEDGMENT

This work was supported by the European Project RTN DRIVEfrom the 6th Framework Program (no.MRTN-CT-2004-512224). The results were generated using CASTEP within the program package Material Studio, developed by Acers Inc., San Diego.

REFERENCES

1. F. Silva, X. Bonnin, J. Achard, O. Brinza, A. Michau and A. Gicquel, *J. Crystal Growth* **310**, 187 (2008).
2. D.G., Goodwin and J.E. Butler, "Theory of diamond chemical vapor deposition", *Handbook of industrial diamonds and diamond films*, ed. M.A. Prelas, G. Popovici and L.K. Bigelow (New York, NY: Marcel Dekker, Inc. 1997) pp. 527–581.
3. C. Wild, R. Kohl, N. Herres, W. Müller-Sebert and P. Koidl, *Diam. Rel. Mater.* 373-381, 3 (1994).
4. R. Locher, C.Wild, N. Herres, D. Behr and P. Koidl, *Appl. Phys. Lett.* **65**, 34 (1994).
5. J. Achard, F. Silva, O. Brinza, A. Tallaire and A. Gicquel, *Diam. Rel. Mater.* **16**, 685 (2007).

6.   Th. Frauenheim, G. Jungnickel, P. Sitch, M. Kaukonen, F. Weich, J. Widany and D. Porezag, *Diam. Rel. Mater.* **7**, 348 (1998).
7.   S. J. Harris and D. G. Goodwin, *J. Phys. Chem.* **97**, 2 (1993).
8.   B. J. Garrison, E. J. Dawnkaski, D. Srivastava and D. W. Brenner, *Science* **255**, 835 (1992).
9.   K. Larsson and J.-O. Carlsson, *Phys. Status Solidi A* **186**, 319 (2001).
10.   T. Van Regemorter and K. Larsson, *J. Phys. Chem. A* **113**, 3274 (2009).
11.   T. Van Regemorter and K. Larsson, *Diam. Rel. Mater.* **18**, 1152 (2009).
12.   T. Van Regemorter and K. Larsson, *J. Phys. Chem. A* **112**, 5429 (2008).
13.   P. Hohenberg, P. And W. Kohn, *Physical Review B* **136**, *136* (1964).
14.   W. Kohn, W. And L. Sham, *Phys. Rev. A* **140**, 1133 (1964).
15.   J. Perdew, and Y. Wang, *Phys. Rev. B* **45**, 13244 (1992).
16.   P. Ziesche, S. Kurth, and J. Perdew, *Computational Materials Science* **11**, 122 (1998).
17.   J. Perdew, J. Chevary, S. Vosko, K. Jackson, M. Pederson, D. Singh and C. Fiolhais, *Phys. Rev. B* **46**, 6671 (1992).
18.   M. Teter, M. Payne, and D. Allan, *Phys. Revs B* **40**, 12255 (1989).
19.   H. Monkhorst, and J. Pack, *Phys. Rev. B* **13**, 5188 (1976).
20.   T. Fischer, and J. Almlof, *J. Phys. Chem* **96**, 9768 (1992).
21.   M. Segall, R. Shah, C. Pickard and M. Payne, *Phys. Rev, B* **54**, 23 (1996).
22.   K. Larsson and J.-O. Carlsson, *Phys. Rev. B* **59**, 8315 (1999).
23.   T. Van Regemorter and K. Larsson, *Chem. Vap. Dep.* **14**, 224 (2008).

Mater. Res. Soc. Symp. Proc. Vol. 1203 © 2010 Materials Research Society   1203-J17-27

Electrospray deposition of diamond nanoparticle nucleation layers for subsequent CVD diamond growth

Oliver JL Fox, James OP Holloway, Gareth M Fuge, Paul W May, Michael NR Ashfold
School of Chemistry, University of Bristol, Bristol, BS8 1TS, U.K.

ABSTRACT

Nucleation is the rate-determining step in the initial stages of most chemical vapour deposition processes. In order to achieve uniform deposition of diamond thin films it is necessary to seed non-diamond substrates. Here we discuss a simple electrospray deposition technique for application of 5 nm diamond seed particles onto substrates of various sizes. The influence of selected parameters, such as experimental spatial arrangement and colloidal properties, are analysed in optimizing the method by optical and electron microscopy, both before and after nanocrystalline diamond deposition on the seed layer. The advantages and limitations of the electrospray method are highlighted in relation to other commonly exploited nucleation techniques.

INTRODUCTION

The nucleation of non-diamond substrates prior to chemical vapour deposition (CVD) of nanocrystalline (NCD) and ultra-nanocrystalline diamond (UNCD) films is crucial for creating smooth, homogenous and uniform thin films. Substrates can be treated to promote surface nucleation in a variety of ways [1-3] including mechanical or ultrasonic abrasion [4-5], bias-enhanced nucleation [6-7] and diamond-containing-photoresist coating [8].

Mechanical abrasion can be efficient at nucleating substrates for microcrystalline diamond (MCD) deposition, but the low surface roughness of NCD and UNCD films often demands a less aggressive nucleation technique. In previous studies, it has been shown that spraying diamond seed particles directly onto substrates can achieve nucleation densities of around 10^{11} cm^{-2} [9-10]. The electrostatic spray ("electrospray") deposition technique uses a large potential difference to ionize nucleating particles (or droplets containing the particles) via the corona effect and accelerates them towards a grounded substrate. The technology has found many applications in the materials industry for application of particulates, paints and powders onto a variety of complex substrate forms and materials [11]. The technique has been employed successfully to seed silicon substrates prior to diamond CVD [12-13] using diamond powders suspended in water. In this case, the particles were forced through drying media (removing the majority of the liquid) by application of 30 psi pressure into an electrostatic spray gun (with an ionizing electrode at 80 kV) where they became charged and were then applied to the substrate.

Here we describe a variation on this simple yet versatile electrospray deposition method for coating substrates with a nucleation layer comprised of 5 nm diamond particles. The main difference is that both the dispersion medium and the diamond particles exit through the electrode nozzle but the volatile liquid then evaporates before reaching the substrate, resulting in a uniform but dense coating of diamond particles. The potential difference pulls the suspension through the capillary, removing the need to apply a driving pressure, while charging prevents the droplets coalescing.

Subsequent NCD or MCD growth by microwave (MW) plasma enhanced CVD allowed optimization of this method, indicating where seeding uniformity could be improved. Scaling up the technique allowed uniform coverage of planar substrates with diameters up to 2 inches and provided a novel means of seeding more complex three-dimensional substrates.

EXPERIMENT

The application of colloidal particles onto a substrate surface requires only a simple experimental setup which is detailed in Fig.1. Approximately 1 ml of the colloidal suspension (i.e. nanodiamond particles in the chosen dispersion medium) is placed in a plastic syringe located on the outside of an insulating box. The 21G syringe needle (outside diameter = 0.81 mm) passes through the wall of the box by turning through 90° and once inside the box bends through an angle θ to point upwards. The angle θ was varied during the optimization of the apparatus to achieve the best substrate coverage. A 35 kV bias was applied to the metal tip of the nozzle which was sufficient to ionize the droplets without becoming unstable and arcing. The substrate (silicon, unless otherwise stated) was positioned ~50 mm away from the nozzle tip on a moveable, conducting mount which could be adjusted through three axes of freedom. The mount was well grounded, ensuring that ionized aerosols emanating from the nozzle would be attracted along electric field lines towards the substrate.

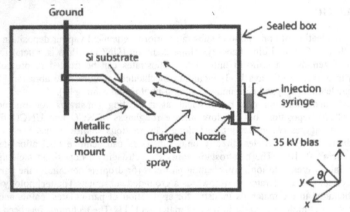

Figure 1. Schematic diagram detailing the electrostatic spray deposition apparatus. Coordinates are used to define the angle of the nozzle relative to the substrate.

When using substrates with diameters greater than ~10 mm the seeding layer was found to be significantly improved by spinning the substrate, thus combining two techniques commonly utilized in the materials industry. Rotating the substrates on the mount at up to 1500 r.p.m. enabled uniform layers to be deposited onto substrates up to 2 inches in diameter. The deposition took just 3 minutes to complete, but could also be repeated to achieve a thicker layer of seed particles if required, although this risked clustering seeds on top of each other.

After seeding, NCD or MCD films were grown at Bristol using an Astex-type 2.45 GHz MWCVD reactor, and by Advanced Diamond Technologies Ltd., USA (ADT) by hot filament

(HF) CVD to get an indication of the nucleation density and uniformity of the subsequent film coverage. These diamond thin films were analyzed by scanning electron (JEOL JSM 6330F) and optical microscopy, and with laser Raman spectroscopy at room temperature and excitation wavelengths of both 325 nm (UV, HeCd) and 514 nm (green, Ar^+) using a Renishaw 2000 spectrometer.

DISCUSSION

In addition to the experimental hardware described above, the following two parameters were investigated to optimize the coverage of diamond nanoparticle seed layers on Si substrates. These were assessed using optical and electron microscopy both before and after growth of CVD diamond films.

Spatial arrangement of nozzle and substrate

The dimensions of the substrate determine its positioning within the electrospray apparatus. Small substrates, less than 10 mm in diameter, could be mounted on the underside of a plate positioned in the xy plane (as defined in Fig.1) but ~20 mm higher (i.e. along z) than the nozzle tip. For larger substrates, and those requiring simultaneous spin coating, a larger mount was prepared as shown in Fig.1 with the surface to be coated at $\theta = 50°$.

Similarly, the orientation of the nozzle in relation to the substrate was also a key factor in guaranteeing an even coating of nanodiamond. The optimal angle was found to be $\theta = 90°$ (i.e. the nozzle pointing along the z axis and, therefore, at 40° to the substrate surface). At angles θ <90°, residual 'splash marks' were visible on the substrate where the liquid had not completely evaporated in flight. This highlights another variable: the nozzle-to-substrate distance, which was adjusted to be sufficiently large that all the liquid evaporated from the droplets before reaching the substrate.

Properties of the colloidal suspension of nanodiamond particles

The successful transfer of nanodiamond to, and coating of, substrates depended primarily on the physical properties, such as viscosity, dielectric constant and surface tension, of the colloidal suspension exiting from the nozzle. Studies with similar nozzles have suggested that the primary droplets were generally monodisperse with sizes in the order of 1 μm although this is strongly dependent on the flow rate through the nozzle [11]. Increasing the concentration of the suspension had the effect of reducing the droplet size while also inducing agglomeration of particles. The presence of such agglomerates on the substrate increased the surface roughness of the subsequent film.

Maintaining a suspension of diamond nanoparticles with a narrow size distribution was vital for the homogenous coating of the substrates and depended on treatment of both the raw material and the suspension. The nanodiamond used in this study was formed by detonation synthesis in an oxygen-deficient atmosphere and had the smallest sizes (4 to 5 nm) of all synthetic particulate diamond [14-15]. These nanoparticles, as supplied, are clustered into tightly bound secondary agglomerates which dramatically increase the average particle size and need to be broken up before being dispersed into the suspension for electrospraying. Recent advances have enabled de-agglomeration of these secondary particles to produce clear, monodisperse nanodiamond colloids. The technique, outlined by Ozawa *et al.* [16], combines two processes (stirred-media-milling using zirconia beads and high-power ultrasonication) into a method termed bead-assisted sonic disintegration (BASD). The BASD technique achieves the required ~5 nm particulate size in about 30 minutes and is, therefore, much quicker than previous bead milling techniques. The presence of agglomerates sprayed onto the substrates may well be attributable to incomplete dispersion of the nanodiamond particles.

Perhaps the most crucial parameter in the electrospray techique described here is the liquid used to disperse the nanodiamond. Several criteria influence the choice of dispersion media. The boiling point and viscosity of the liquid determines how quickly it evaporates from the droplet when travelling between the nozzle and the substrate. Generally, a low boiling point/low viscosity liquid, such as methanol, is favoured because this should help ensure only the diamond nanoparticles deposit on the substrate. Using more viscous liquids than methanol, such as water and cyclohexane, causes unwanted splashing of liquid onto the substrate, leaving 'drying marks' in the seed layer. In the case of cyclohexane, the nozzle-to-substrate distance can be increased to counteract this problem. The high surface tension forces of water also have to be overcome as the droplets emerge from the nozzle tip. Related to this property, the polarity of OH bonds in water and alcohol molecules can be used to stabilize the charged nanodiamond particles. In this regard, water is the best liquid for nanodiamond particles but, as shown previously [13], it is not suited to the electrospray apparatus unless the liquid is removed from the particle stream before the ionization stage. Combining sufficient bond polarity and low boiling point, we found that alcohols provide the most suitable dispersion media for the deagglomerated nanodiamond particles. The OH bond polarity allows formation of hydrogen bonds to the charged diamond particles thereby decreasing the propensity to flocculation.

A variety of liquids were tested, including isopropyl alcohol, ethanol, methanol, cyclohexane and deionised water. The optimal bond polarity and viscosity was achieved using methanol, which became the dispersion medium of choice in this experimental setup. In many other cases, the addition of diamond nanoparticles to the liquid resulted in spontaneous flocculation into large clusters of particles, as revealed by the change from a clear to cloudy colloidal suspension.

To stabilize the 5 nm diamond particles in liquids less polar than water, we investigated addition of a polymer to coat the primary particles and prevent their aggregation. Polyvinylpyrrolidone (PVP) is sufficiently polar, due to the carboxyl group on the pyrrole ring, to adsorb onto the charged surface of the nanodiamond particles. This generates a steric layer that prevents the particles getting sufficiently close to agglomerate via van der Waals forces. However, despite successfully dispersing the colloid in almost all the liquids (except for the non-polar cyclohexane where significant flocculation was observed), the use of PVP polymer was abandoned following analysis of the seed layers by optical microscopy and of the grown films by electron microscopy. The polymer increased the viscosity of the droplets and was also found to transfer to the substrate. As this polymer layer dried it formed a cracked pattern on the substrate which was still visible in the diamond films post-growth. It was thought that the extreme conditions of the MWCVD plasma might reduce the influence of this patterning. To some extent this was true, as the nanocrystalline films deposited in the microwave reactor were more uniform than the UNCD films grown in the hot-filament reactors at ADT.

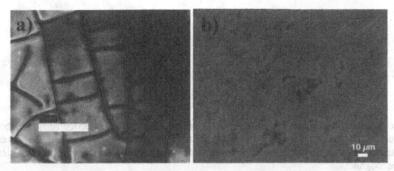

Figure 2. Patterning of substrates due to the PVP polymer drying and cracking, observed a) prior to diamond deposition by optical microscopy (×20) and b) following HFCVD of a UNCD layer by electron microscopy (used with permission from ADT, Ltd.)

The patterning can be seen in Fig.2a, which shows a seeded film prior to diamond deposition. The severity of this problem precludes the use of PVP to stabilize the suspension, limiting the types of liquid which can be used in electrospray deposition. De-agglomeration of the detonation nanodiamond by BASD in methanol, the dispersion medium to be used in this electrospray technique, minimized flocculation and thus removed the need for polymer stabilization.

Analysis of films following diamond deposition

The success of a nucleation layer is best established by growing a thin film, and two methods were employed in this study. The original optimization of the electrospray apparatus involved growing a nanocrystalline diamond layer by MWCVD using an argon rich plasma ($95\%Ar/1\%CH_4/4\%H_2$) and later testing by growing UNCD films on 2-inch Si wafers at ADT by HFCVD. The latter study enabled comparison with ADT's standard ultrasonic nucleation technique. An example of an UNCD film grown by HFCVD on a nanodiamond/PVP coated substrate is shown in Fig.2b.

The cracked patterning due to the dried PVP polymer is still present in the HFCVD grown UNCD film, whereas NCD films grown at Bristol by MWCVD using the same seeding technique are uniform and do not exhibit the lined pattern (Fig.3a). The difference is likely to be due to two important factors. First, the ADT films are much thinner (1.15-1.25 μm for the UNCD films compared with >5 μm for the Bristol NCD films), so inhomogeneities in the seed layer are more likely to be observed in the UNCD layer. Secondly, the higher growth rates, substrate temperatures and atomic hydrogen density associated with the MWCVD technique will reduce the influence of non-uniform areas of the seed layer and 'burn off' any remaining polymer.

Analysis of the surface roughness of HFCVD grown UNCD films using a stylus profilometer over a 2 mm scan shows considerably higher values for the electrospray seeded films compared to the ADT seeding method. However, Raman analysis (not shown) confirms that the quality of the diamond (the ratio of sp^3-to-sp^2 carbon content) is not influenced by the nucleation method chosen having similar spectra for both electrospray and ADT seeded films.

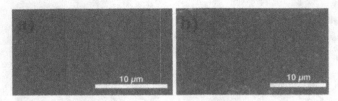

Figure 3. NCD films grown by MWCVD on a) a nanodiamond/PVP seeded substrate showing good uniformity and no evidence of polymer influencing the diamond layer, and b) a diamond film seeded with a PVP-free nanodiamond/methanol dispersion – showing almost identical surface roughness.

After seeding with the BASD nanodiamond in methanol, removing the need for PVP stabilization, another set of NCD films were grown by MWCVD and an example is shown in Fig.3b (UNCD deposition has not been carried out on these substrates at the present time). The NCD films grown with and without PVP in the seed layer display very similar surface roughness.

A major advantage of this electrospray technique over photoresist spin-coating or other abrasive seeding methods is the ability to nucleate more complex shapes other than planar substrates. To this end, we have demonstrated successful seeding of a variety of three dimensional substrates, including 10 mm^3 tungsten carbide cubes and patterned Si wafers (examples of which are shown in Fig.4a and 4b).

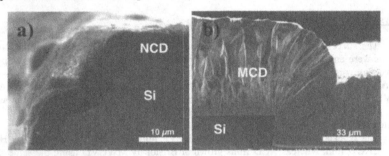

Figure 4. Examples of electrospray seeding using nanodiamond in methanol and subsequent diamond growth on patterned silicon substrates: a) section through thick NCD film and Si substrate showing successful seeding and subsequent growth around substrate edge and b) section of thick MCD film on patterned Si substrate containing recessed areas.

CONCLUSIONS

We have demonstrated a quick, simple and versatile electrospray technique for seeding substrates with a nucleating layer of nanodiamond particles. The addition of a colloid stabilizing polymer in the seed suspension was investigated but rejected as it can cause unwanted patterning of the seed layer which is particularly apparent in thin UNCD films grown by HFCVD. De-agglomerating the nanoparticles by the BASD method in methanol removed the need for addition of any polymer and allowed the colloid to be directly sprayed onto a range of substrate sizes and materials. This seed layer has been shown to be very suitable for subsequent growth of uniform, homogenous films of both nano- and microcrystalline diamond. Any modification of the nanodiamond structure due to the experimental parameters has not been studied but appears not to influence the seed layer or subsequent film growth whereas it could be problematic for more diverse applications.

ACKNOWLEDGMENTS

The authors thank Prof. Anke Krüger (University of Würzburg, Germany) and Dr. Masaki Ozawa (Meijo University, Japan) for kindly supplying samples of their BASD nanodiamond in various liquids, Dr. Nicolaie Moldovan (ADT, Ltd.) for growth of UNCD films, Element Six Ltd. for the long term loan of the MW reactors used in this work, and EPSRC and the Royal Society Wolfson Laboratory Refurbishment Scheme for funding.

REFERENCES

1. D. Das, R. N. Singh, *International Materials Reviews* **52**, 29 (2007) and references therein.
2. S-T. Lee, Z. Lin, X. Jiang, *Materials Science and Engineering* **25**, 123 (1999).
3. Y.K. Liu, P.L. Tso, I.N. Lin, Y. Tzeng, Y.C. Chen, *Diamond Rel. Mater.* **15**, 234 (2006).
4. A. A. Morrish, P. E. Pehrsson, *Appl. Phys. Lett.* **59**, 417 (1991).
5. L. Demuynck, J. C. Arnault, C. Speisser, R. Polini, F. LeNormand, *Diamond Relat. Mater.* **6**, 235 (1997).
6. M. J. Chiang, M. H. Hon, *Diamond Relat. Mater.* **10**, 1470 (2001).
7. M. D. Irwin, C. G. Pantano, P. Gluche, E. Kohn, *Appl. Phys. Lett.* **71**, 716 (1997).
8. G. S. Yang, M. Aslam, *Appl. Phys. Lett.* **66**, 311 (1995).
9. Y. Tang, D. M. Aslam, *J. Vac. Sci. Technol. B* **23**, 1088 (2005).
10. G. S. Yang, M. Aslam, K. P. Kuo, D. K. Reinhard, J. Asmussen, *J. Vac. Sci. Technol. B* **13**, 1030 (1995).
11. Y. Matsushima, Y. Nemeto, T. Yamazaki, K. Maeda, T. Suzuki, *Sensors and Actuators B* **96**, 133 (2003).
12. A. P. Malshe, R. A. Beera, A. A. Khanolkar, W. D. Brown, and H. A. Naseem, *Diamond Relat. Mater.* **6**, 430 (1997).
13. A. P. Malshe, A. A. Khanolkar, W. D. Brown, S. N. Yedave, H. A. Naseem, M. S. Haque *Proceedings Of The Fifth International Symposium On Diamond Materials, Electrochemical Society Inc.* **97**, 399 (1998).
14. J. B. Donnet, C. Lemoigne, T. K. Wang, C. M. Peng, M. Samirant, A. Eckhardt, *Bull. Soc. Chem. Fr.* **134**, 875 (1997).
15. O. A. Shenderova, V. V. Zhirnov, D. W. Brenner, *CRC Crit. Rev. Solid State Mater. Sci.* **27**, 227 (2002).
16. M. Ozawa, M. Inaguma, M. Takahashi, F. Kataoka, A. Krüger, E. Osawa, *Adv. Mater.* **19**, 1201 (2007).

Mater. Res. Soc. Symp. Proc. Vol. 1203 © 2010 Materials Research Society   1203-J17-33

Hydrogen Incorporation in MPCVD Nanocrystalline Diamond During the Deposition Process

Dominique Ballutaud[1], Marie-Amandine Pinault[1], François Jomard[1], Alain Lusson[1] and Samuel Saada[2]

[1]CNRS-GEMaC, 1 place Aristide Briand, 92195 Meudon cedex, France
[2]CEA, LIST, Diamond Sensor Laboratory, 91191 Gif-sur-Yvette, France

ABSTRACT

Hydrogen incorporation is studied in two Microwave Plasma CVD nanocrystalline diamond films deposited with prolonged bias or not during the growth step. The hydrogen content and bonding forms are analysed by Secondary Ion Mass Spectrometry, Raman and Fourier Transformer Infrared Spectroscopy. Our results show a high hydrogen concentration up to 3.10^{21} cm^{-1}, as expected in nanocrystalline diamond, and in good agreement with the sp^2 phase rate measured by Raman spectroscopy . The FTIR spectra exhibit two sharp peaks at 2850 and 2920 cm^{-1} and show that a fraction of hydrogen is bonded to sp^3 CH$_2$ groups. Hydrogen desorption experiments are performed to analyse the local structure modification of the diamond films.

INTRODUCTION

High intrinsic hydrogen concentrations - up to 10^{21} cm^{-3} for nanocrystalline diamond (NCD) - are found in diamond grown by microwave plasma chemical vapour deposition (MP CVD)[1]. The presence of disordered hydrogenated sp^2 carbon in grain boundaries, a high density of dislocations which may trap hydrogen, together with the presence of H$_2$* dimers, are supposed to explain this high hydrogen concentration [2–4]. Bulk hydrogen is stable in crystalline diamond up to 950 °C [5]. The NCD films structure and the related properties depend on the growth conditions, on the grain size but also the crystalline quality of the diamond phase [6]. In this work, we study the effects of a prolonged bias applied during the growth step on the hydrogen concentration and bonding forms in the diamond films. Hydrogen concentration is measured by secondary ion mass spectrometry (SIMS) and the local structure of the diamond films is analysed par Raman and Fourier Transformed Infra-red Spectroscopy (FTIR). Out-diffusion of bulk hydrogen is performed by thermal annealing under ultra-high vacuum and the local structure modifications are analysed.

EXPERIMENTAL

The MPCVD reactor used for the NCD film deposition was described previously [7]. Two different deposition processes were used : the sample A was grown with a methane concentration of 10%, with a bias enhanced nucleation; the sample B was grown with the same methane concentration and same parameters, but with a bias voltage of −307 V applied during all the growth step and a biasing current of 58 mA. Figure 1 show the SEM micrographs of the sample [7]. For sample A, the size of the grain is about 9 nm. Sample B exhibits a different particular morphology with clustering of the nanograins at the top of the film. The films have the same thickness of 200 nm. The hydrogen concentration profiles were analysed with a CAMECA IMS 4f secondary ion mass spectrometer with Cs^+ primary ion beam. Quantifications of hydrogen were achieved by using CVD diamond standards implanted with a known dose of hydrogen, with a 10% error in accuracy. Raman spectroscopy measurements were performed using a Jobin-Yvon/Horiba HR800 coupled with an 800 mm focal length spectrograph, equipped with a liquid nitrogen cooled CCD detector. The He–Cd laser line at 325 nm was used as excitation source (resolution 1.5 cm^{-1}). Infrared absorption experiments have been performed at 10 K with a BOMEM DA8 Fourier transform interferometer with a resolution of 0.5 cm^{-1}. The as-grown samples were submitted to annealing under ultra-high vacuum (10^{-10} Torr) at 1050 °C during 8h.

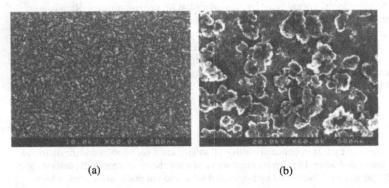

(a)   (b)

Figure 1. SEM micrographs : (a) sample (A) ; (b) sample B deposited with bias during all the growth.

RESULTS AND DISCUSSION

Hydrogen content in as-grown and annealed samples.

The SIMS hydrogen concentration profiles of the as-grown and annealed samples are reported on figure 2. Hydrogen concentration is higher in sample B deposited under bias. After the thermal treatment (8h at 1050 °C) performed on the two samples, some hydrogen is still present in the diamond films. Hydrogen is not completely out diffused. This result means that either the hydrogen bonds are very stable, or the diffusion of hydrogen in the nanocrystalline diamond films is very low. In case of H_2 molecules presence, they should dissociate in atomic hydrogen to outdiffuse.

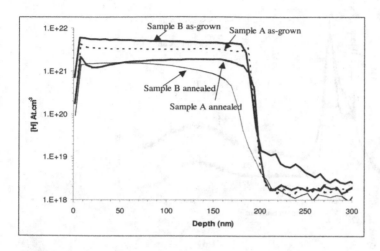

Figure 2. Hydrogen concentration profiles of samples A and B, as-grown and after thermal annealing.

Ultra-violet Raman spectroscopy

Figure 3 (a and b) shows the evolution of the characteristic line of the diamond (1332.5 cm^{-1}), of the two D and G bands (respectively1350 and 1550 cm^{-1}), and of the CH$_x$ band (3000 cm^{-1}) of the Raman spectra for the two samples. The sample A spectrum (figure 3a) exhibits the diamond line and the D and G bands attributed respectively to C=C sp^2 bond stretching mode related to the presence of all sp^2 structures including both olefine chains or aromatic rings, and to the collective breathing mode of aromatic rings [8]. Hydrogen concentration and sp^2/sp^3 ratio in sample A are in good agreement with previous results obtained for nanocrystalline samples [9] with similar grain size. The shape of the D and G bands are modified and the 3000 cm^{-1} CH$_x$ band intensity has decreased when this sample is annealed. In sample B (figure 3b), the 1332.5 cm^{-1} diamond line cannot be detected, as it merges as a shoulder into the D band. After thermal annealing of sample B, the D and G band have been modified, the CH$_x$ band intensity has decreased, and the diamond characteristic line is then visible. A shoulder is also observed at about 1170 cm^{-1}, assigned to transpolyacetylène segments at grain boudaries [10], which has disappeared after thermal annealing.

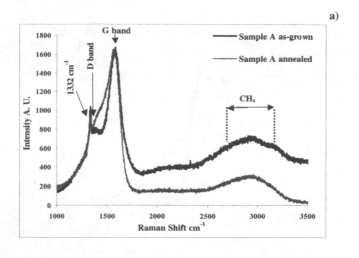

a)

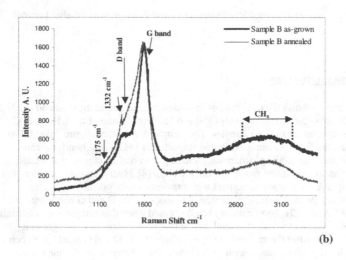

(b)

Figure 3. Raman spectra of samples A and B as-grown and after thermal annealing: (a) samples A ; (b) sample B.

FTIR spectroscopy

Figure 4 (a and b) shows the FTIR spectra of samples A and B recorded at 10 K. Beside the different total hydrogen content in these films, it may be observed that hydrogen is bonded to different CH_x (x=1,2,3) groups ([11]. The FTIR spectra show a broad band with two sharp peaks at 2846 and 2920 cm^{-1} which may be attributed to sp^3 CH_2 groups. The large band at

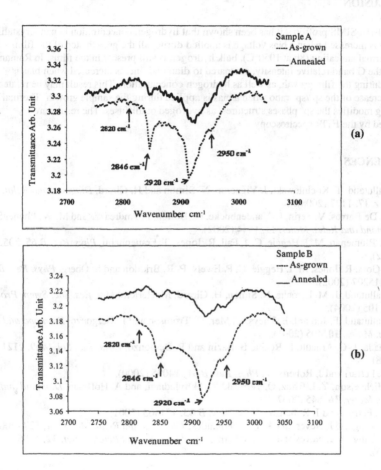

Figure 4. FTIR spectra of the as-grown samples and after thermal annealing: (a) sample A; (b) sample B.

2880 cm^{-1} is associated with sp^3 CH$_3$. A shoulder at 2820 cm^{-1} is attributed to hydrogen trapped in intragranular defects [12], and is more visible for sample B. After the samples thermal annealing, the total FTIR spectrum intensity has decreased.

CONCLUSION

Using SIMS profiling, it has been shown that hydrogen concentration in nanocrystalline diamond is increased when a bias voltage is applied during all the growth step of the film. After thermal annealing (8h at 1050°C), bulk hydrogen is still present in the films. In Raman spectra, the G band relative intensity, compared to diamond line, is increased when bias is applied during the film growth, as well as hydrogen concentration. This result may be related to the increase of the sp^2/sp^3 ratio when a bias is applied during the sample growth. Thermal annealing modifies the sp^2 phases structures, as hydrogen outdiffuses. The results are confirmed by the FTIR spectroscopy.

REFERENCES

1. D. Ballutaud, T. Kociniewski, J. Vigneron, N. Simon and H. Girard, *Diamond and Relat. Mater.* **17**, 1127 (2008).
2. M. I. De Barros, V. Serin, L. Vandenbucke, G. Botton, P. Andreazza and M. W. Phaneufe, *Diamond and Relat. Mater.* **11**, 1544 (2002).
3. A. T. Blumenau, M. I. Heggie, C. J. Fall, R. Jones, T. Frauenheim, *Phys. Rev. B* **65**, 205205 (2002).
4. J. P. Goss, R. Jones, M. I. Heggie, C. P. Ewels, P. R. Briddon and S. Öberg, *Phys. Rev. B* **65**, 115207 (2002).
5. D. Ballutaud, J.-M. Laroche, N. Simon, H. Girard, M. Herlem, *Mat. Res. Soc. Symp. Proc.* **813**, 105 (2004).
6. D. Ballutaud, F. Jomard, B. Theys, C. Mer, D. Tromson and P. Bergonzo, *Diamond and Relat. Mater.* **10**, 405 (2001).
7. S. Saada, J.-C. Arnault, L. Rocha, B. Bazin and P. Bergonzo, *Phys. Stat. Sol. (a)* **9**, 2121 (2008).
8. A. C. Ferrari and J. Robertson , *Phys. Rev. B* **61**, 14095 (2000).
9. S. Michaelson, Y. Lifshitz, O. Ternyak, R. Akhvlediani, and A. Hoffman, *Diamond and Relat. Mater.* **16**, 845 (2007).
10. A. C. Ferrari and J. Robertson , *Phys. Rev. B* **63**, 121405 (2001).
11. C. J. Tang, A. J. Neves and A. J. S. Fernandes, *Diamond and Relat. Mater.* **11**, 527 (2002).
12. C. J. Tang, A. J. Neves and A. J. S. Fernandes, *Diamond and Relat. Mater.* **12**, 1488 (2003).

Nanodiamond

Mater. Res. Soc. Symp. Proc. Vol. 1203 © 2010 Materials Research Society   1203-J08-01

Nanodiamond Particles in Electronic and Optical Applications

O. Shenderova[1], S. Ciftan Hens[1], I.Vlasov[2], V. Borjanovic[1], G. McGuire[1]
[1]International Technology Center, Raleigh, NC, 27617 (USA)
[2]General Physics Institute, Russian Academy of Sciences, 119991 Moscow (Russia)

ABSTRACT

Current use of nanodiamond (ND) particles in electronic-related applications is mostly restricted to their role in seeding of substrates for growth of diamond films by chemical vapor deposition (CVD). While it is a niche application, nanometer-sized diamond particles are indispensable in this role. Seeding of substrates using a novel slurry of detonation nanodiamond particles in di-methylsulfoxide (DMSO) and methanol is one of the topics of this article. At the same time, optical applications of NDs, particularly development of photoluminescent NDs for biomedical applications is one of the most popular current research topics. In this paper perspectives for the use of detonation NDs and specifically the role of surface functionalization in imparting photoluminescent properties to detonation NDs as well as enhanced photoluminescence of proton irradiated ND-polydimethylsiloxane composites are discussed.

INTRODUCTION

While the importance of high quality diamond films in optical applications and electrically conductive doped diamond films in electronics and bioelectronics is well recognized and they are in high demand, the role of nanodiamond (ND) particles in these applications is not so obvious. So far ND in the particulate form plays a secondary role in diamond-related electronics and this role is mostly based on its small size, such as, for example, in seeding of substrates for CVD diamond film growth [1]. Another application illustrating the use of ND in electronic application is related to electron field emission, where a coating of detonation NDs (DNDs) particles on Si or metal tips can improve cold cathode performance due to an overall decrease of the work function of the structure [2]. While the listed applications are rather narrow the range of applications in electronics can be very broad if electrically conductive nanodiamond particles can be synthesized. Production of conductive boron-doped nanodiamond particles can be very beneficial and can find broad applications in high surface area carbon electrodes for electroanalysis, electrochemical double-layer capacitors, storage materials for batteries, electrocatalyst support material for fuel cells, stationary support for liquid chromatography and other applications [3].

Growth of diamond films on a foreign substrate using a CVD process typically requires methods to enhance the nucleation of diamond, such as bias-enhanced nucleation or the scratching techniques. One of the popular method is seeding of the substrate with diamond particles through the use of a slurry of diamond particles dispersed in an appropriate solvent accompanied by ultrasonic agitation. Detonation ND particles, produced from carbon-containing explosives, have an average primary particle size of only 4-5 nm with a narrow size distribution that seems ideal for growth of diamond films. However DNDs have a serious drawback – they form tightly bound agglomerates of primary particles during synthesis and purification. While these large agglomerates were acceptable for growth of microcrystalline diamond films, a new generation of nanocrystalline and ultrananocrystalline diamond films intended for use as thin membranes in bioapplications and complicated patterns in MEMS and NEMS, for example, demands a new generation of DND seeds for their growth. We will discuss recent advances in DND deagglomeration and fractionation that has resulted in DND slurries with more consistent and predictable properties. In a recent paper we discussed the advantages of using DMSO for DND fractionation as well as the surface chemistry requirements of DND to be efficiently processed in DMSO. We also presented seeding results using different seeding slurry compositions [4]. In the current paper we will summarize findings on the role of DMSO in fractionation of DND and discuss more in depth what properties of DMSO and other solvents are important for DND dispersion and seeding.

DNDs may have high surface charges (both positive and negative) in a variety of solvents, depending upon the types of functional groups on its surface [5]. This allows the manipulation of charged DND particles in a suspension using an electric field. There are many examples whereby electromanipulation of nanoparticles play important roles, such as in the concentration of colloidal particles from solution, nanoparticle separation, transport, formation of coatings on a substrate, and the formation of micropatterns for biosensor fabrication. For example, the combined dielectrophoretic/electrophoretic deposition of DND has been used to coat microscopic silicon tips with nanodiamond for enhanced electron emission in cold cathode applications [2]. Electrophoretic seeding of substrates with submicrometer diamond particles for the growth of diamond films has been explored [6]. The electrophoretic collection of DND probes for attachment of proteins was explored by Hens et al. [7]. In this work, it was found that the electrophoretic collection of DND on a silicon substrate was dependent upon both the field strength and the time of electrophoresis. The substrate saturates rapidly showing a concomitant drop in current from the electrode. This method was used for the collection of target-bound NDs, whereby the target streptavidin was collected using a ND-biotin probe, by applying a potential across an electrically conductive substrate containing a field tip array (FTA) [7]. It is anticipated that the electrophoretic and dielectrophoretic manipulation of nanodiamond will afford many applications for trapping, manipulation and separation of biomolecules. In addition, because of the high surface charge of DND and its relatively high dielectric constant, DND movement in the electric field can be controlled with high precision by properly designing the electrode configuration [8].

Nanodiamonds posses a variety of very attractive optical properties, such as transparency (in UV-VIS-IR spectral range), high index of refraction useful due to high scattering of light for ND visualization inside cells [9], pronounced Raman signatures for ND detection [10] and others. Based on some of these properties, application of DNDs in protection from UV radiation (sunscreens, cosmetics, plastics, paints, etc) was suggested [11]. Photonic crystals formed from monodispersed nanoparticles have found applications in bio-chemical sensors. Recently it was demonstrated that colloids of irregularly shaped, highly charged detonation nanodiamond particles with low monodispersity can form colored structures and planar photonic crystals exhibiting exceptionally intense iridescence [12]. These findings suggest that, if mechanically stabilized, the use of nanodiamond-based photonic devices could offer numerous benefits based on the optical transmission of diamond in the UV–VIS–IR spectral regions, its high refractive index, ease of surface functionalization and robustness of the diamond particles and related composites.

One of the unique physical properties of nanodiamond, also characteristic of bulk diamond, is its intrinsic photoluminescence originating from structural defects and impurities (dopants). Of particular importance for the intrinsic PL properties of diamond are N-related defects (complexes of substitutional N atoms and vacancies). It is well recognized that photoluminescent NDs are highly desirable as biological labels and traceable molecular carriers, as well as stable alternatives to fluorescent "dyes" for special applications. Unlike molecular dyes, photoluminescent NDs do not photobleach. This results in a fluorescent signal of superior intensity and stability. Unlike molecular dyes, PL NDs do not photoblink thus they have been traced in time using rapid sampling in a cell's 3D space. An additional advantage over molecular dyes is that a wider variety of biomolecules can be attached to NDs.

In general, DND is not considered as a perspective material for intrinsic PL emission based upon the creation of N-V centers [13]. Chemical modification of the DND surface, however, can enhance photoluminescence. Several DND surface functionalizations performed in our lab have demonstrated that it is possible to enhance the PL intensity of the original DND powder by an order of magnitude by attaching small molecules to its surface [14]. These small molecules have no intrinsic PL characteristics of their own. In similar work, it was recently reported by Mochalin et al. that the chemical modification of DND surfaces with octadecylamine resulted in a highly fluorescent material [15]. Interestingly, it was found that proton irradiation of DND embedded into the polymer polydimethylsiloxane (PDMS) resulted in a strong dose-dependent photoluminescence. PDMS without DND and pure DND powder did not show the same effect, even at higher irradiation doses. It was suggested that this photoluminescence may arise from defects formed at the interface of a DND/PDMS matrix [16].

In the current paper several applications of nanodiamond particles of detonation synthesis out of those listed above will be discussed in more detail. Particularly, we will discuss using DND for seeding of substrates for CVD diamond growth (electronics-related application) as well as photoluminescent properties of DND induced by functionalization and irradiation of ND-PDMS composites.

EXPERIMENTAL DETAILS

The polydispersed DNDs produced using TNT and RDX explosives by the detonation synthesis method have average primary particle sizes of 4 nm. Diamond samples were supplied by New Technologies Co. (Chelyabinsk, Russia), the Alit Inc (Ukraine), The Institute of Biophysics, SB RAS (Krasnoyarsk, Russia), and the Real Dzerzinsk (Russia). DNDs possessing a positive zeta potential were purified using a solution of chromic anhydride in sulfuric acid; a mixture of $NaOH/H_2O_2$ and ion-exchange resins. DNDs possessing negative zeta potential were purified by annealing in air at 425°C for 1 hour [17]; in a fluidized-bed type reactor supplied with ozonized air at elevated temperature [18], by annealing in air after being mixed with boric anhydride; treatment in saturated basic media using atomic oxygen. One of the DNDs sample was also annealed in hydrogen atmosphere at 400°C for PL studies.

DND powder was dispersed in a solvent using a custom made or a Cole-Parmer direct-immersion horn-type ultrasound sonicator with an output power of 100 W. DND particle size distributions in suspensions were measured by photon correlation spectroscopy (PCS) using a Beckman-Coulter N5 submicron particle size analyzer. Zeta-potential measurements were collected using a Malvern Instrument Zetasizer Nano ZS.

DND functionalization was performed with silicone-containing groups aminopropyl-triethoxysilane or APES $(Si(OC_2H_5)_3(CH_2)_3NH_2$, diphenyl dichlorosilane $(SiCl_2(C_6H_5)_2)$ and 10-(carboxymethoxy)decyldimethylchlorosilane $(Si(CH_3)_2C_{10}COOCH_3)$ denoted as ND-APES, ND-Ph and ND-SiC10, respectively. The condensation reaction of DND functionalized with hydroxy groups was performed in a solution of toluene. The ND-OH was suspended in dry toluene and the silane functional group was added dropwise into the solution. After addition, the DND was washed with toluene by pelleting to remove unfunctionalized silane.

The PDMS-DND nanocomposite was prepared as reported previously [16]. Samples of pure PDMS and PDMS-DND composite were irradiated in vacuum with a 2 MeV proton beam that was delivered by the 1.0 MV Tandetron accelerator at the Rudjer Bošković Institute, Zagreb, Croatia . Samples were irradiated at a fluence 10^{15} protons/cm^2.

The PL spectra of the samples were recorded at room temperature with a LabRam HR-800 spectrometer using the 488 nm mode of an Ar$^+$ laser. In PL measurements of functionalized DND the laser beam of ~ 0.1 mW power was focused to a spot of 2-3 μm diameter on the thick (~1mm) layer surface of the powder-like samples. In PL measurements of PDMS-DND composites the laser beam of ~ 0.01 mW power was focused to a spot of 10 μm diameter on a polymer film surface.

RESULTS AND DISCUSSION

In the first section, the use of DND for seeding of substrates related to electronic applications of DND will be discussed. Then, the photoluminescent properties of DND induced by functionalization and PL induced by proton irradiation of PDMS-DND composites will be reported.

Seeding Slurries Based on Detonation Nanodiamond in DMSO

Recently, we reported on the role of nanodiamond surface charge (zeta potential) for dispersion and fractionation of DND in water and DMSO [4]. The zeta potential of DND in different solvents characterizes their colloidal stability, whereby the values outside the region +30 to -30 mV indicate good resistance to agglomeration and sedimentation of DND in suspensions. Results of the comparative study of dispersion of 9 samples of DND with positive or negative zeta potentials in DMSO and water are summarized in Fig.1. Samples 1-5 of the studied DNDs have positive Z-potentials, while samples 6-9 have negative Z-potentials in water (Fig. 1). Samples with negative Z-potential have majority of the carboxylic groups, while samples with positive Z-potentials posses hydroxyls, ketons and possibly pyrones groups responsible for negative or positive surface charge of NDs, correspondingly, due to dissociation/protonation of the surface groups. Samples 1 and 3-5 formed relatively stable suspensions in water while sample 2 (poorly purified from metallic impurities and amorphous carbon) quickly sedimented. All samples with negative zeta potential (6-9), on the contrary, formed very stable suspensions in water. As it follows from Fig. 1, NDs dispersed in DMSO have Z-potential signs similar to those in water, but with different values. This figure also shows that DNDs with positive Z-potential in water (high or low in absolute value) have even higher or relatively similar positive Z-potentials in DMSO for the samples tested. Conversely, for DNDs with a negative charge, the zeta potential dramatically decreases in absolute value when DNDs are suspended in DMSO as compared to water. Fig. 1 shows that for DND with negative zeta potentials, the average particle size in DMSO is larger than in water. Thus, it was concluded that DNDs with positive Z-potentials can be successfully dispersed in a DMSO suspension, while it was not possible to reliably disperse DNDs with negative Z-potentials in DMSO.

For a variety of applications, including seeding, suspensions of ND in alcohols are beneficial since alcohols possess low surface tension, low viscosity, and low temperature of vaporization (Table 1). However, alcohol suspensions of polydispersed DND are poorly stabilized even with positive zeta potentials, as compared to the same DND suspensions in water and DMSO (Fig. 2a,b). It was also observed that the fractionation is not efficient for DND suspensions in pure alcohol solvents suspensions (Fig. 2c,d). We explored the use mixtures of solvents with DMSO, whereby the co-solvent was, for example, water, alcohol, or acetone. We reasoned that DMSO may improve the stability of DND suspensions in the co-solvent. We found that DMSO improved the dispersivity and yield of small-sized DND fractions, as compared to their fractionation without the addition of DMSO. The particle sizes and concentrations of DND in the supernatant are summarized in Fig. 2 (c,d). It can be seen that the yield of the small fractions is increased when DMSO is mixed with water or alcohol as compared to using these solvents without the addition of DMSO.

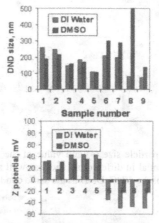

Figure 1. Summary of the comparative study of average agglomerate sizes and values of zeta potentials for 9 DND samples in DI water and DMSO. Samples 1-5 have positive zeta potentials, while samples 6-9 are negatively charged in DI water and DMSO. The average particle sizes are based on intensity-based measurements for the unimodal particle size distribution.

Powders of DND are more economical and safer to transport than suspensions and powder samples allow for flexibility in formulating different types of suspensions. However, preparing DND powders is challenging since drying DND slurries from fractionated suspensions in water causes agglomeration that may increase the DND average size (by volume) by up to 100% (Fig. 3). Luckily, DMSO solutions of DND can be dried and reconstituted (in either DMSO or water) without agglomeration (Fig. 3). Also, alcohol solvents mixed with DMSO can also be dried and reconstituted with only a 50% increase in particle size (Fig. 3).

Previously, we ran a series of DND seeding experiments using 4" Si wafers [4]. Sonic treatment of the wafers in slurries containing 30 nm DND was performed with pure DMSO as well as mixtures of DMSO with methanol using different proportions of solvents. The results of seeding were inspected by SEM. We found that seeding using nanodiamond suspensions in pure DMSO produced many aggregates and was independent of the nanodiamond concentration. This result from using DMSO alone is expected because of DMSO's high surface tension, high boiling point and slow evaporation at normal atmospheric pressures. Thus, the pure DMSO suspensions were considered inadequate for seeding applications. However, excellent colloidal DND seeding suspensions were found for slurries using mixtures of DMSO and methanol, using either

1:1 to 1:4 DMSO:MeOH mixtures. These slurries have been used for seeding of substrates for CVD diamond film growth. Representative example of the seeded substrate is illustrated in Fig. 4.

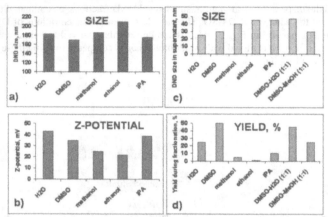

a) b) c) d)

Figure 2. Comparison of the particle size (a) and value of the zeta potential (b) for the DND sample with positive zeta potential in different solvents as well as size (c) and yield in a supernatant (d) after centrifugation of the 1wt% DND at the same conditions in different solvents.

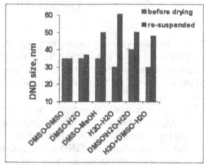

Figure 3. Comparison of DND sizes for the initial suspensions of different solvents and their mixtures as well as after drying and re-suspension, whereby the designation solvent1-solvent2 denotes that the ND solution was dried from solvent 1 and resuspended in solvent 2. For the experiment indicated as H2O + DMSO/H2O, first the nanodiamond was fractionated in water and DMSO was added to the suspension and second the dried DND was resuspended in water.

An important finding of the previous work [4] is the limitation of using NDs that have only positive zeta potentials for DMSO suspensions, such as for dispersions and fractionation. So far, some general speculations on the DND-solvent interaction can be drawn as is shown from the schematics in Fig. 5. DMSO solvent contains a highly polar S=O group, forming very strong hydrogen bonds with molecules possessing acidic hydrogen atoms such as, for example, hydroxyl groups on the surface of DNDs with a positive Z-potential. Possibly the carbonyl Lewis Base on the particle may also interact with the partially positive S of DMSO. In addition to a highly polar group, DMSO contains two hydrophobic CH_3 groups. Thus a solvation shell preventing DND particles from agglomerating may be formed through hydrogen bonds between surface groups of nanoparticles and S=O groups of DMSO enclosed by methyl groups at the periphery of the shell. As a result, there are no 'bridges' between DND particles because the methyl groups form a hydrophobic "shell" around these particles (Fig. 5). DMSO is a very strong electron-donor (Table 1), and its interaction with positively charged DND surface groups (electron acceptor) is stronger than that with negatively charged DNDs. The Lewis base property of DMSO on the donor number scale exceeds the donor numbers for solvents of dimethylformamide (DMF), dimethylacetamide (DMA), and N-methylpyrrolidone (NMP) (data not shown), which are other good aprotic solvents as well as donor numbers of water and alcohols (Table 1). Its acidic number (Lewis Acid properties) exceeds those for DMF, DMA, and NMP, but is 2-3 times lower than that for alcohols and water [19] (Table 1).

Figure 4. SEM image of a part of 4" Si wafer seeded using slurry of 20-30nm DND in DMSO/MeOH (1:1) mixture. Scale bar corresponds to 100nm.

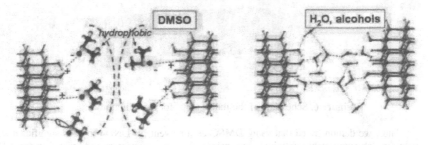

Figure 5. Schematic illustration showing the reason why DMSO is the best solvent for DND with positive zeta potentials as compared to protic solvents. Solvation shells formed by DMSO molecules prevent agglomeration in the solvent and during drying (left), while "bridges" formed by intermolecular hydrogen bonds through water and alcohol molecules (right) promote agglomeration during drying from protic solvents.

For alcohol and water suspensions, we have the opposite effect: 'bridges' are formed between neighboring DND particles, which occurs through hydrogen bonding between the solvent molecules and the ND surface groups as well as solvent-solvent interactions (Fig.5). Drying of NDs from protic solvents, which form hydrogen bond-type bridges between the ND particles, results in irreversible agglomeration during drying. Thus, the striking difference between DMSO and protic solvents is the observation that NDs in DMSO exhibit high dispersivity and an ability to be dried and resuspended without agglomeration.

Although DND powders dried from DMSO prevent agglomeration after resuspension, the process of seeding a substrate with DND requires an alcohol for best results [4]. DND seeds used for CVD diamond film growth are formed on a substrate during ultrasonic treatment possibly by accelerating these nanoparticles toward the substrate through the collapse of cavitation bubbles (Fig. 6, step 1). This causes the nanoparticles to adhere strongly to the solid surface (Fig. 6, step 2). Next, there is the important step of removing the slurry and drying of the substrate (Fig. 6, step 3). The final structures of the particles dispersed over a substrate was observed by SEM in [4], can be 3-D or 2-D build ups of the particles, or a uniform, monodispersed layer (Fig. 6), depending upon the solvent composition. 3-D agglomerate build ups are formed when the volatility of a solvent is low (as for DMSO alone) and thus DND particles "sediment" on top of each other. 2-D agglomerates are possibly formed when the surface tension of a solvent is still too high and H-bond bridges hold particles together. Thus, there are several factors influencing the uniformity of seeding and the prevention of particle clumping during drying; these factors include the solvent's surface tension, volatility, density, viscosity (Table 1) and other factors that still require thorough study. For example, the high surface tension of DMSO needs to be adjusted by adding the co-solvent alcohol to prevent DND build ups.

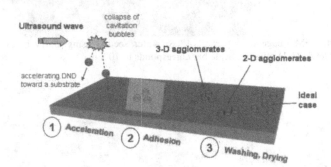

Figure 6. Schematic of the major steps for a seeding process.

Thus, we demonstrated that using DMSO as a solvent for DND provides significant advantages in achieving stable colloidal suspensions that are resistant to sedimentation for NDs with positive zeta potentials. The sizes of ND aggregates dispersed in DMSO are smaller than agglomerates seen for the same NDs suspended in DI water. Using DMSO allows for the effective fractionation of a variety of DNDs including those that cannot be fractionated, for example, using DI water as a solvent. In addition, ND slurries in DMSO can be mixed with a variety of other polar solvents, expanding the range of applications of the slurries, such as substrate seeding for CVD diamond growth. In addition, ND slurries in DMSO may be useful for biological applications since this solvent is used in pharmacology, medicine, and biotechnology.

Table 1. Physico-chemical characteristics of the solvents used in the study for DND fractionation, drying and resuspension as well as for seeding by ultrasonic treatment of a substrate in a DND slurry.

solvent	Dielectric constant	Dipole moment	Donor number (Lewis base)	Acceptor number (Lewis acid)	Density, g/m^3	Interfacial Tension, mJ/m^2	Vapor pressure, kPa 25°C	Viscosity, cP
DMSO	46.68	3.96	29.8	19.3	1.1	42.98	0.1 (20°C)	2.14
H₂O	78.46	1.85	18.0	54.8	1.0	71.81	3.2	0.894
Methanol	32.70	1.70	19.0	41.5	0.791	22.30	17	0.544
Ethanol	24.55	1.69	19.2	37.9	0.789	21.90	7.9	1.074
IPA	20.33	1.58	19.8	37.3	0.785	23.30	6.1	1.96
Role in seeding	Colloidal stability; prevention of agglomeration due to hydrogen bonded network during drying				Cavitation, sonochemistry, prevention of agglomerates build up during washing & drying			

Photoluminescent properties of functionalized DND and DND-PDMS composite

Of particular importance for the intrinsic PL properties of diamond are N-related defects (complexes of substitutional N atoms and vacancies). While large portion of the nanodiamond particles produced by HPHT method can be converted to PL nanoparticles [20], formation of optical centers in majority of detonation ND particles based on N complexes with vacancies is still elusive. Recently Turner et al. [21] probed individual particles of the detonation nanodiamond sample and unambiguously proved that N is located in the core of these particles using STEM-EELS analysis. Model-based quantification indicated a nitrogen content of around 3 atomic percent in all samples. Thus these EELS measurements indicate that nitrogen impurities, which are deemed to contribute to the photoluminescent effect in DND samples, are embedded with a tetrahedral coordination (sp³) within its core and are not positioned close to the particle surface or within the graphitic shell. More over, recently it was demonstrated that small portion of DND particles with primary particle sizes more than 20-30nm can be converted to PL particles based on N complexes with vacancies [22]. Larger particles contain extended defect-free parts of a crystal, where substitutional N can be preserved and active N-V centers formed, while proximity of defects and surface acceptors in smaller particles may result in suppression of the optical activity of N-V centers in DND. This finding suggest that through careful control of N content in the precursor material as well as its incorporation into DND core during synthesis, PL DND, in principle, can be produced.

Alternatively, chemical modification of the DND surface can enhance broad-band photoluminescence commonly observed in this material [14, 15]. Several DND surface functionalizations performed in our lab have demonstrated that it is possible to enhance the PL intensity by an order of magnitude as compared to the original DND powder by attaching small molecules to their surface, whereby these small molecules do not have intrinsic PL characteristics of their own [14]. In the current work we demonstrate increased PL by attaching organic molecules through silicon bridges to DND surface. Furthermore, PL studies of DND containing Si atoms within functional groups was motivated by recent observations by Borjanovic et al. [16] on significantly enhanced photoluminescence of polydimethilsiloxane PDMS-DND composite samples induced by high energy proton irradiation.

The PL of all studied functionalized DND photobleached under laser light, decreasing in intensity by factors 2-5 (depending on the samples) for the first 500 s of laser irradiation, as shown in Fig. 7 for the ND-APES sample. To observe the PL photobleaching effect we used the same laser irradiation as for PL excitation. Stabilized PL spectra of functionalized DNDs and initial DND used for functionalization (measured at 10 min after starting of laser irradiation) are illustrated in Fig. 8. PL intensities in their maxima (about 560 nm) are increased by factors 2, 4, 5, and 20, respectively, for the samples ND-H, ND-SiC10, ND-Ph, and ND-APES in comparison to the PL intensity of original non-functionalized DND. Enhancement of PL intensity for the ND-APES sample is very significant and useful for applications. Note an appearance of a narrow line at around 700 nm in the PL spectra of ND-Ph and ND-APES samples (Fig.8). The study of its origin is in progress, but it can be hypothesized that defected silica domains responsible for enhanced photoluminescence [23] are formed on DND surface in a case of DND functionalization with aminopropyltriethoxysilane as well as during proton irradiation of PDMS-DND composite containing alternating Si and O atoms within a back-bone of PDMS.

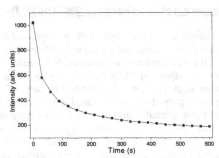

Figure 7. Dependence of 560 nm PL intensity on the time duration of laser irradiation of ND-APES sample.

We recently reported [16] the phenomenon of high-energy proton-induced and fluence-dependant photoluminescence of PDMS and PDMS-DND (2% by weight) nanocomposite material. Pure PDMS exhibits blue PL, while the PDMS-DND composites exhibit a pronounced stable green PL under an excitation light of 425 nm. It was observed that the PL of PDMS-DND composite is much more prominent than that of pure PDMS, as well as that for pure DND powder, even when DND powder was irradiated with a fluence that was one order of magnitude greater. Thus, it is reasonable to argue that the origin of enhanced PL for the proton irradiated PDMS-DND composite may be attributed to a combination of 1) enhanced intrinsic PL within ND particles due to proton implantation generated defects and 2) by PL that originates from structural transformations produced by protons at the nanodiamond/matrix interface [16]. Here we present the PL experimental results on high energy proton irradiation of a 1 wt% PDMS-DND nanocomposite material as compared to a pure PDMS material, which were irradiated under the same conditions. The PL spectra for original and irradiated PDMS and PDMS-DND samples are presented in Fig. 9. In the analysis of the intensity, the difference in sample transparency and the estimated efficient PL/Raman probe volume was taken into account. It can be estimated from Raman signal ratio that efficient PL/Raman probe volume is 5 times larger for pure transparent PDMS than for PDMS-DND nanocomposite. There is a three fold difference in the intensities of PDMS-DND as compared to the pure irradiated PDMS. Thus, it can be concluded that presence of 1wt% of DND shows a total enhancement of approximately 15 times as compared to irradiated pure PDMS for this irradiation fluence.

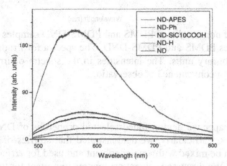

Figure 8. PL spectra measured in 10 min after starting of laser irradiation for the pristine ND powder (black line, control), DND treated in hydrogen (orange line, ND-H) at 600°C and 3 types of functionalized NDs: ND-APES (Red line), ND-Ph (green line), ND-SiC10 (blue line, denoted in figure as ND-SiC10COOH). Excitation wavelength is 488 nm.

Thus, this study strongly supports PL enhancement phenomenon that is induced by high energy proton irradiation of PDMS and PDMS-DND samples. It is also shown that the PL enhancement is not as significant as it was reported in [16] where the PL emission intensity of the PDMS-DND composite was approximately a factor of 25 higher than for the pure PDMS. However, the samples of PDMS-DND nanocomposite material in the current study were prepared with a 2 fold lower DND content. It can be concluded that the PL intensity enhancement depends on the DND content in the nanocomposite material. This study and the previously reported study [16] indicate the importance of the nanodiamond/matrix interface as well as the structural transformations, which are produced by protons at the nanodiamond/matrix interface.

Detailed experimental Raman and FTIR data and analysis of vibrational spectra for irradiated PDMS and DND-PDMS samples were recently reported in [24]. The findings showed that DND-PDMS nanocomposite materials exhibited enhanced stability against high energy proton irradiation as compared to pure PDMS, which may allow for additional perspectives for the use of DND as a filler in DND-based polymer composites for applications in high radiation environments.

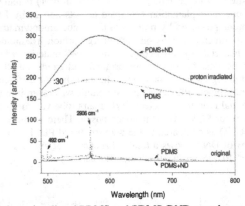

Figure 9. PL spectra of non-irradiated PDMS and PDMS-DND samples compared with PL spectra of irradiated samples PDMS and PDMS-DND. The spectra for irradiated samples are shifted in intensity by 150 arbitrary units. The intensities in the spectra of irradiated samples are decreased by 30 times for a convenience of observation.

CONCLUSIONS

In the current paper we demonstrated selected properties of DND useful for electronics and optical applications. Particularly, we demonstrated that slurries of DND with positive zeta potential in DMSO can be mixed with a polar solvent and used for efficient substrate seeding for CVD diamond growth. We discussed solvent parameters important for this DND application.

Considering optical applications of DND, we showed that attaching organic molecules through the silicon-containing bridges to DND surface as well as embedding the DND in the Si-based polymer matrix followed by proton irradiation treatment have significant influence on photoluminescent properties of the produced complexes. Functionalized DND as well as DND embedded in PDMS polymer matrix and irradiated by high energy proton beam demonstrated greatly enhanced photoluminescence in comparison with PL of pristine DND powder or irradiated pure PDMS. These findings emphasize the role of presence of Si atoms/complexes at the DND surface/interface with DND. This phenomenon definitely requires further study, but it can be hypothesized on the role of formed defected silica domains responsible for enhanced photoluminescence.

Acknowledgment
The author would like to thank the U.S. Army Research Laboratory for their support under grant W911NF-04-2-0023. Dr. M. Jaksic and I. Zamboni for assistance with proton irradiation of samples. Dr. G. Cunnigham for silicone-related functionalization. Dr. T. Feygelson, NRL, for treatment of DND in hydrogen.

REFERENCES

1. Williams, O.A., Douheret, O., Daenen, M., Haenen, K., Osawa, E., and Takahashi, M. (2007) *Chem. Phys. Let.* 445, 255-258.
2. Alimova, A.N., Chubun, N.N., Belobrov, P.I., Detkov, P.Y., and Zhirnov, V.V. (1999) *J. Vac. Sci. Tech. B* 17, 715-718.
3. A. Ay, V. M. Swope, G. M. Swain, *J. Electrochem. Soc.*, (2008), 155, 10, B1013-B1022.
4. Shenderova, O., Hens, S.C., *Diam. Relat. Mat.* In press.
5. Schrand, A. M., Hens, S.C., Shenderova, O. A. *Crit. Rev. Sol. St. Mat. Sci.* (2009), 34, 18.
6. Zhitomirsky, I. (1998) *Mater. Lett.* 37, 72-78.
7. Hens, S.C., Cunningham, G., Tyler, T., Moseenkov, S., Kuznetsov, V., and Shenderova, O. (2008) *Diam. Relat. Mat.* 17, 1858-1866.
8. Hughes, M.P. (2000) *Nanotechnology* 11, 124-132.
9. Smith, B.R., Niebert, M., Plakhotnik, T., and Zvyagin, A.V., (2007) *J. Lumines.*, 127, 260.
10. Perevedentseva, E., Cheng, C.Y., Chung, P.H., Tu, J.S., Hsieh, Y.H., and Cheng, C.L., (2007) *Nanotechnology*, 18, 315102.
11. Shenderova, O., Grichko, V., Hens, S., and Walsh, J. (2007) *Diam. Relat. Mat.* 16, 2003.
12. Grichko V., Tyler T., Grishko V., Shenderova O. (2008) *Nanotechnology* 19, 225201.
13. J-P. Boudou, P. A Curmi, F. Jelezko, J. Wrachtrup, P. Auber; M. Sennour, G. Balasubramanian, R. Reuter, A. Thorel, E. Gaffet, (2009) *Nanotechnology 20*, 235602.
14. Shenderova, O., Hens, S.C., chapter 4 in book (2010) *Nanodiamonds: Applications Toward Biology and Nanoscale Medicine*, Springer-Verlag, ed. by Dean Ho.
15. V. N. Mochalin, and Y. Gogotsi, (2009) *J. Am. Chem. Soc. 131*, 4594-4595.
16. Borjanovic, V., Lawrence, W.G., Hens, S., Jaksic, M., Zamboni, I., Edson, C., Vlasov, V., Shenderova, O., and McGuire, G. (2008) *Nanotechnology*,19,45, 455701.
17. Shenderova, O., Petrov, I., Walsh, J., Grichko, V., Grishko, V., Tyler, T., and Cunningham, G. (2006) *Diam. Relat. Mat.* 15, 1799-1803.
18. Petrov, I., Shenderova, O., Grishko, V., Grichko, V., Tyler, T., Cunningham, G., and McGuire, G. (2007) *Diam. Relat. Mat.* 16, 2098-2103.
19. W.R.Fawcett, (2004) *Liquids, Solutions, and Interfaces*, Oxford.
20. Y.-R. Chang, H.-Y. Lee, K. Chen, C.-C. Chang, D.-S. Tsai, C.-C. Fu, T.-S. Lim, Y.-K. Tzeng, C.-Y. Fang, C.-C. Han, H.-C. Chang, W. Fann, *Nat. Nanotechnol.* 2008, *3*, 284-288.
21. Turner, O, I. Lebedev, I., Shenderova, O., Vlasov, I., Veerbeck, J., Tendeloo, G. Van (2009) *Adv. Funct. Mater.* 19, 1-9.
22. I. Vlasov, O. Shenderova, S. Turner, O. I. Lebedev, A.A. Basov, I. Sildos, A. Shiryaev and G. Van Tendeloo, *Small*, 2009, in press
23. Cai, K. F., Zhang, A. X., Yin, J. L., Wang, H. F., Yuan, X. H. (2008) *Appl. Phys. A* 91, 579–584.
24. Borjanović, V., Bistričić, L., Vlasov, I., Furić, K., Zamboni, I., Jakšić, M., Shenderova, O. (2009) *J. Vac. Sci. Tech. B* 27, 6, 2396-2403.

Mater. Res. Soc. Symp. Proc. Vol. 1203 © 2010 Materials Research Society   1203-J03-05

Fluorescent Nanodiamonds: Effect of Surface Termination

I. Kratochvílová[1], A. Taylor[1], A. Kovalenko[1], F. Fendrych[1], V. Řezáčová[1], V. Petrák[1], S. Záliš[2], J. Šebera[2], M. Nesládek[3]

[1]Institute of Physics, Academy of Sciences Czech Republic v.v.i, Na Slovance 2, CZ-182 21, Prague 8, Czech Republic

[2]J. Heyrovský Institute of Physical Chemistry, AS CR, v.v.i., Dolejškova 3, CZ-182 23, Prague 8, Czech Republic

[3]Hasselt University, Institute for Materials Research (IMO), Wetenschapspark 1, B-3590 Diepenbeek, Belgium

ABSTRACT

It has been reported that physico-chemical properties of diamond surfaces are closely related to the surface chemisorbed species on the surface. Hydrogen chemisorption on a chemical vapor deposition grown diamond surface is well-known to be important for stabilizing diamond surface structures with sp^3 hybridization. It has been suggested that an H-chemisorbed structure is necessary to provide a negative electron affinity condition on the diamond surfaces. Negative electron affinity condition could change to a positive electron affinity by oxidation of the H-chemisorbed diamond surfaces. Oxidized diamond surfaces usually show characteristics completely different from those of the H-chemisorbed diamond surfaces. The unique electron affinity condition, or the surface potential, is strongly related to the chemisorbed species on diamond surfaces. The relationship between the surface chemisorption structure and the surface electrical properties, such as the surface potential of the diamond, has been modelled using DFT based calculations.

INTRODUCTION

Nanodiamond is a novel promising material for *in-vitro* and *in-vivo* imaging in living cells. Specially, nanocrcystalline diamond (ND) and ultrananocrytslaine diamond (UND) particels [1-2] offer novel advantages for the drug delivery development [3]. Advantage of nano diamond is also the ability to penetrate into the cells without evoking a toxic cell response. The newly developed production, purification and functionalisation techniques enable the material to be widely used. The adsorption strength on nanodiamond due to hydrophilic and hydrophobic interactions is so high that a very efficient capture of proteins such as cytochrome c, myoglobin and albumin occurs. Recent studies have also initiated the use of ND/UNDs for in-vivo molecular imaging and bio-labelling. Additionally nanodiamond's offer properties such as the biocompatibility, nontoxicity and a possibility of the lable-free imaging, based on easily detected Raman signal and intrinsic fluorescence from nanoparticle point defect - Nitrogen-vacancy centres. In diamond the association of a vacancy with a nitrogen impurity leads to the formation of a luminescent defect, called the NV colour centre and being either neutral (NV^0) or negatively charged (NV^-) [4,5]. Both of these centres are photostable and can be detected at the individual level which allows application of nanodiamonds for functional intracellular imaging on the molecular level based on tagging specific molecular sites. This photoluminescence can be altered by many parameters - applying a magnetic field, electric field, microwave radiation and also by diamond surface termination.

It has been reported that physico-chemical properties of diamond surfaces are closely related to the surface chemisorbed species on the surface [6]. Hydrogen chemisorption on a chemical vapor deposition -grown diamond surface is well-known to be important for stabilizing diamond surface structures with sp^3 hybridization. Many reports have suggested that an H-chemisorbed structure is necessary to provide a negative electron affinity condition on the diamond surfaces. It was reported that the negative electron affinity condition could change to a positive electron affinity by oxidation of the H-chemisorbed diamond surfaces. Oxidized diamond surfaces usually show characteristics completely different from those of the H-chemisorbed diamond surfaces. The unique electron affinity condition, or the surface potential, is strongly related to the chemisorbed species on diamond surfaces [7]. The differences between O-terminated and H-terminated diamond surfaces include electrical conductivity, hydrophobicity and hydrophilicity, negative and positive electron affinities and fluorescence. In this work we study the influence nanodiamond termination on its NV-centers fluorescence. We have combined Raman spectroscopy and theoretical approaches (quantum chemical calculation) to get deeper insight into the complex behaviour and properties of variously terminated diamond nanoparticle with NV centers.

PHOTOLUMINESCENCE

ND prepared from detonation synthesis contains significant amounts of sp^2 carbon and are therefore not efficient emitters of N-V related luminescence. For this reason we used novel techniques to prepare high pressure high temperature (HPHT) ND by specific milling techniques [8]. Hydrogen and carbonyl terminated ND were obtained from synthetic ND of 20-50 nm size (Fig. 1). ND was first oxidised to produce carbonyl termination. The ND were then exposed to an H-plasma (2500 W, 600 °C, duration 2-5 minutes) in AX6500 SEKI reactor to obtain hydrogen-terminated ND. Finally, the H terminated diamond was annealed 120 minutes in air at 500^0C to revert the surface to carbonyl termination. All the in ND termination states were confirmed by ATR-FTIR and XPS.

Raman (514 nm excitation wavelength) spectra (Fig. 1) were taken from untreated ND, carbonyl terminated ND, H terminated ND, and annealed H terminated ND using a Renishaw InVia Raman Microscope at 80K. Measurements show that for untreated ND and carbonyl terminated ND NV^0 and NV^- related luminescence can be observed at 575 nm and 637 nm respectively. H terminated ND show no NV related luminescence under the same experimental conditions. However, after annealing H terminated ND the luminescence related to NV centres was again observed. This phenomenon (luminescence depending on the ND termination) can be used for the detection of chemical processes ongoing on the surface.

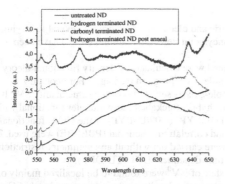

Fig.1: Raman and Photo Luminescence spectra taken from untreated ND, carbonyl terminated ND, H terminated ND, and annealed H terminated ND.

ATOMIC FORCE MICROSCOPY

The topography of the nanodiamond particles was investigated by atomic force microscopy (AFM). The AFM measurements were performed with NTEGRA Prima NT MDT system (Ireland) under ambient conditions [9]. The tip - sample surface interaction monitored the van der Waals force between the tip and the surface; this may be either the short range repulsive force (in contact-mode) or the longer range attractive force (in non-contact - tapping mode). The AFM depicted nanodiamond (Fig. 1) particles were performed using the tapping mode. The samples were scanned under the soft CSG 10 type of probe. The tapping mode consisted of oscillating the cantilever at its resonance frequency (14-28 kHz) and lightly "tapping" the tip on the surface during scanning. Fig 2 shows AFM visualised nanodiamond particles.

Fig. 2: AFM depiction of nanodiamond particles on mica surface. NTEGRA Prima NT MDT system (Ireland).

85

COMPUTER EXPERIMENTS

Quantum chemical calculations of geometric and electronic structure for several low-lying excited states of the neutral nitrogen-vacancy point defect in diamond (NV^0) have been performed using finite model clusters.

Electronic structures of all systems examined were calculated by density functional theory (DFT) methods with the Gaussian 09 and Turbomole-5.10. program packages. For geometry optimization, either 6-31G* polarized double-ζ basis sets (G09) or the same quality SVP basis (Turbomole) for H, C, N, atoms. Either the hybrid Becke's three-parameter functional with the Lee, Yang, and Parr correlation functional (B3LYP) (G09/B3LYP) or the GGA functional using Perdew, Burke, and Ernzerhof exchange and correlation functional (PBEPBE) were used. The geometry optimizations and calculations were carried out without any symmetry restriction. Electronic excitations were calculated by time dependent DFT (TD-DFT).

Unpaired electrons in the ground doublet state of NV^0 were found to be localized mainly on three carbon atoms around the vacancy and the electronic density on the nitrogen and rest of C atoms is only weakly disturbed.

The electronic and structural factors that may impact the effect of luminiscence of ND nanomaterials were investigated by DFT calculations. In order to get deeper insight into the electronic structure of ND particles containing NV centers, nanodiamond clusters of different size and multiplicity, terminated by hydrogen, carbonyl and hydroxyl groups were modeled. NV^0 centers were modeled by replacing carbon atoms in neighborhood of vacancy by nitrogen according to the procedure described in [10, 11]. Clusters $C_{35}H_{36}$, $C_{84}H_{64}$ and their derivatives containing one NV^0 center terminated either by hydrogen or CO and OH groups were optimized using DFT methodology for ^2A and ^4A states.

The influence of the hydrogen substitution by O and OH on doublet excitation energies was calculated by TD-DFT for different models derived from $C_{33}N_1VH_{36}$ (selected structures are depicted in Figure 3). The substitution influences the calculated excitation energies from the ^2A ground state and ^2A - ^4A separation. Calculated lowest lying excitation energies of substituted ND are lowering with the degree of the substitution (e.g. 2.24 eV calculated for $C_{33}N_1VH_{36}$ vs. 2.08 eV for $C_{33}N_1VH_{30}O_6$), which can enable to observe the luminescence for substituted ND. Calculated doublet - quadruplet separation also diminish by the substitution, going from 0.22 eV calculated for $C_{33}N_1VH_{36}$.

The impact of the hydrogen substitution by O and OH on the NV^0 luminiscence probability can be summarized as:

- Carbonyl groups substitution influences the **shapes of orbitals** involved in excitations
- **Excitation energies** shift to lower values due to the carbonyl termination - enlarge luminiscence probability
- **Excitation energies** diminish with the degree of substitution-enlarge luminiscence probability.

Fig. 3: The optimized geometries of hydrogen terminated (left), hydrogen and oxygen terminated (middle) and totally substituted (right) ND clusters derived from $C_{33}N_1VH_{36}$

Acknowledgment

This work was supported by the Academy of Sciences of the Czech Republic Grant Nos., KAN401770651, KAN200100801, KAN400480701 and KAN100400702, the Czech Science Foundation 203/08/1594, the Ministry of Education of the Czech Republic (OC 137), and institutional funds of the Institute of Physics AS CR, v.v.i. AV0Z10100520.

REFERENCES

1.   S. Koizumi, Ch. Nebel, M. Nesladek :Physics and Applications of CVD diamond. Wiley-VCH Verlag GmbH&Co KGaA, Weinheim (2008)
2.   Ristein J., Nesladek M., Haenen K Microscopic diagnostics of DNA molecules on mono-crystalline diamond, Phys. Stat. Sol. A, 204, pp.2835-2835 (2007)
3.   Shu-Jung Yu,Ming-Wei Kang, Huan-Cheng Chang, Kuan-Ming Chen, Yueh-Chung Yu, J. Am. Chem. Soc, 127, pp. 17604-17605 (2005)
4.   K. Iakoubovskii, G.J. Adriaenssens, M. Nesladek, J. Phys. C, 12, 189-199   (2000)
5.   B. Marczewska, P. Olko, M. Nesladek, MPR Waligorski, Y. Kerremans, Rad. Prot. Dosimetry 101, pp 485-488 (2002)
6.   Minoru Tachiki, Yu Kaibara, Yu Sumikawa, Masatsugu Shigeno , Hirohumi Kanazawa, Tokishige Banno, Kwang Soup Song, Hitoshi Umezawa, Hiroshi Kawarada, Surface Science vol. 581, pp. 207–212 (2005)
7.   F. Maier, M. Riedel, B. Mantel, J. Ristein, L. Ley, Phys. Rev. Lett. 85, p. 3472 (2000).
8.   A. Krueger, Chemistry, 14, pp. 1382-1390 (2008)
9.   I. Kratochvilova, K. Kral, M. Buncek, A. Viskova, S. Nespurek, A. Kochalska, T. Todorciuc, M. Weiter, B. Schneider, Biophysical Chemistry, 138, pp. 3-10 (2008)
10.   A. S. Zyubin, A. M. Mebel, M. Hayashi, H. C. Chang, S. H. Lin, J. Comput. Chem. 30, pp. 119-131 (2009)
11.   A. S. Zyubin, A. M. Mebel, M. Hayashi, H. C. Chang, and S. H. Lin, J. Phys. Chem. C , 113, pp. 10432–10440 (2009)

Mater. Res. Soc. Symp. Proc. Vol. 1203 © 2010 Materials Research Society   1203-J17-16

Nanodiamonds Particles as Additives in Lubricants

M.G. Ivanov[1], S.V. Pavlyshko[2], D.M. Ivanov[1], I. Petrov[3], G. McGuire[4], O. Shenderova[4]

[1]Ural State Technical University, Mira Str. 19, Yekaterinburg, 620002, Russia
[2]Institute of Engineering, Science Ural Branch RAS, Komsomolskaya St. 34, Yekaterinburg, 620219, Russia
[3]SKN, Snezinsk, Mira St. 26-97, Snezinsk, 456776, Russia
[4]International Technology Center, 8100 Brownleigh Dr., Raleigh, NC 27617, U.S.A.

ABSTRACT

In the current work we report results of tribological testing of stable colloidal dispersions of detonation nanodiamond (DND) and polytetrafluoroethylene (PTFE) in mineral oil based greases as well as in polyalphaolefin (PAO) oil. Testing has been performed on these formulations using ring-on-ring, shaft/bushing and four ball test techniques. The test results demonstrated significant improvement for tribological characteristics (friction coefficient, extreme pressure failure load and diameter of wear spot) for certain formulations. A strong synergistic effect when using a combination of DND/PTFE additives was observed by a sharp decrease of friction coefficient. It was also demonstrated that using DND with smaller aggregate size (10nm versus 150nm) resulted in better lubricating performance of PAO-based composition.

INTRODUCTION

Within the last decade, certain nanomaterials in powder and colloidal forms have emerged as potential anti-friction and anti-wear additives to a variety of base lubricants [1]. Detonation nanodiamonds with 4-5 nm primary particle size and spherical shape [2] seem to be among the most promising candidates as additives to such lubricants. Until recently, detonation soot, which is a mixture of graphite and nanodiamond particles, has been the predominant material of this class used as an additive to lubricants [3]. It was assumed that the combination of nanodiamond and the graphite phase is beneficial since graphite on its own has lubricating property, while DND contributes by polishing of asperities on friction surfaces and thus reducing friction. Highly purified DND with small aggregate sizes are, however, a relatively new nanomaterial additive of interest for more demanding tribological applications [8-11]. For a long time, it was assumed that pure DND (without sp^2 coating as opposite to DND in the soot) was not suitable for lubrication, because of the abrasive nature of diamond particles. However works by Ivanov et al. [8-10] and Puzyr et al. [11] demonstrated successful application of pure DND for enhanced tribological performance of greases and oils. Ivanov et al. [8-10] also demonstrated that a combination of DND with ultradispersed PTFE particles provides excellent lubrication properties in mineral oils of Class I and greases, well exceeding those for soot and when using only PTFE additives. It is speculated that this combination creates a robust tribological film (PTFE-DND) that can withstand higher loads and can better protect friction surfaces from failure.

In the current work, we further extend these earlier studies[8-10] to greases of different composition as well as report first results of a DND/PTFE composition with PAO as a base oil. Previously polydispersed diamond with average aggregate sizes about 200-250nm were used in experiments [8-10]. With the development of novel methods for deagglomeration of tight diamond aggregates or fractionation of polydispersed material, even much greater improvements in their friction and wear performance is expected. In the current work DND with fraction sizes 150nm and 10nm had been used for preparation PAO oil based compositions. It was demonstrated that deagglomerated DND added to the PAO-based oil indeed provides better lubricating properties .

EXPERIMENTAL DETAILS

Detonation nanodiamonds (DND) were used as additives to lubricants in the present work. DNDs were synthesized at the high pressure-high temperature conditions achieved within the shock wave resulting from the detonation of carbon-containing explosives with a negative oxygen balance [2]. The primary particle size produced by this method is approximately 3-5nm in the most current commercial products. Importantly, primary nanodiamond particles produced by detonation of carbon containing explosives form tightly bonded aggregates during synthesis and purification. DND samples were obtained from Electrochimpribor, Lesnoy, Ural region (for grease and I12 oil experiments) and from SKN, Snezinsk, Ural region (for PAO oil experiments). The size of DND aggregates obtained from Electrochimpribor was 180-200nm. DNDs used for PAO oil experiments were deagglomerated and fractionated to two fractions with average agglomerate size 150nm and 10nm as measured by the dynamic light scattering technique in water suspensions. Dispersion of the DND in the oil was assisted by ultrasonic treatment.

Several types of ultradispersed PTFE particles were used in combination with DND in the lubrication compositions. PTFE-A and PTFE-B particles were obtained by gaso-thermodynamic extraction of residues of an industrial fluoropolymer condensed into ultradispersed powder [12]. The average particle sizes of PTFE-A and PTFE-B were 0.1-1.0um and 1-2um, correspondingly. PTFE-C with average particle size 5um was obtained by milling of the electron irradiated PTFE powder [12]. While PTFE-A and B have spherical form, PTFE-C material has the form of ribbons. As additives to PAO oil, Zonyl MP 1100 PTFE-COOH with average particle size of 2.0-3.0um produced by DuPont, USA was used.

Soft grease was obtained using mineral oil as the base oil and polyurea or Li-based thickeners. According to the National Lubricating Grease Institute (NLGI) classification, the grade of the soft greases prepared in the present study is 1-2. The mineral oils used in the study, I40A and I12A, according to the classification of industrial oils have viscosity grades ISO VG 68 and ISO VG 15, correspondingly. Table 1 summarizes the compositions of the greases with DND and PTFE additives used in the tribological tests. In some experiments (with I12A oil), an oil-soluble sulfonate and friction modifier (SAS) was used. In another set of experiments, synthetic polyalphaolefin-based oils, PAO-6 produced by ExxonMobil as a trade mark SpectraSyn was used.

Table 1. Compositions of the greases using I40A as a base oil with DND (180-200nm) and PTFE additives used for shaft/bushing and ring-on-ring tests. Amount of additives is shown in wt.%.

	Sample 1	Sample 2	Sample 3	Sample 4	Sample 5	Sample 6
grease	Li-grease	Li-grease	Li-grease	Polyurea-grease	I40A	Polyurea-grease
PTFE	PTFE-B 4%	PTFE-A 4%	PTFE-C 4%	PTFE-C 4%	PTFE-A 4%	PTFE-C 4% PTFEA1.3%
DND	0.13%	0.13%	0.13%	0.3%	1.6%	0.7%

In the experiments below testing was performed on greases and PAO oil -based formulations using ring-on-ring, shaft/bushing (analog - Almen-Wieland-Tribometer) and four ball extreme pressure (EP) tests. Test apparatus SMT-1 was used for the ring-on-ring and the shaft/bushing tests. In the ring-on-ring test, quenched steel rings ШХ-15 (USA analog steel 52100) with hardness HRc=52 and flat friction surfaces with roughness Ra=0.38 were used. The rotational velocity was 500, 1000 and 1500 rpm. Rings were pressed together by a spring with a force of 314N. In the shaft/bushing tests, shaft and bush were made from un-quenched steel and 20XH3A quenched steel, correspondingly. The rotational velocity was 300 rpm. The load was increased in increments of 500N until the failure load was reached. The scar diameter was measured using a standard four-ball technique. Balls made from steel 40X (USA analog steel 5140) with diameter 12.70mm were used. The rotational velocity of the upper ball was 1460rpm and the load was 196N. Time of loading was 60 minutes. EP failure mode in the four-ball test was defined at rotational velocity 1460 rpm and at a load increased with increments 490N applied during 10s for every load value (analog ASTM D 2783-88).

RESULTS AND DISCUSSION

Testing was performed on the above formulations using ring-on-ring (for friction coefficient), shaft/bushing (for friction coefficient and extreme pressure failure load) and four ball tests (forEP failure load and scar diameter in wear measurements). In the first section below, coefficient of friction and EP failure load of several grease formulations are reported for DND and several types of PTFE (A, B, C) measured in shaft/bushing and ring-on-ring tests. Then, results of the friction coefficient obtained in the ring-on-ring test for the oil formulation using I12A as the base oil and DND/PTFE additives are discussed. Finally, tribological test results addressing the role of DND size in the PAO oil based formulations using four ball tests and ring-on-ring techniques are provided.

Tribological properties of the greases with DND and PTFE additives

Results of the shaft/bushing test for the samples 1-3 and 6 (Table 1) are illustrated in Fig.1 for friction coefficient as a function of applied load. As can be seen from Fig.1, friction coefficient is a non-linear function of the applied load in the shaft/bushing test and also depends on the type of PTFE, indicating that PTFE-C (ribbon form) is more suitable for tribological applications. Best performance was observed for sample 6, containing the largest amount of DND (0.7%) in this series of experiments. For this sample, friction coefficient is as low as 0.05 for the whole range of the load (0.5-5kN).

Fig.2 (a) illustrates failure load for the grease samples 1-5 (Table 1) with DND and PTFE additives obtained from the shaft/bushing test. As compared to the failure load of pure Li-grease, 126 kG, all samples essentially demonstrate an increase in the failure load with the best performance for the compositions corresponding to the samples 4 and 6 (PTFE-C and 0.3% and 0.7% of DND, correspondingly). In fact, the maximum load to failure for sample 6 with 0.7% DND exceeded the maximum load limit (i.e., 5kN) for the test apparatus.

Fig.2 (b) demonstrates results on friction coefficients for the ring-on-ring test at different rotation speeds and low load (320N), as compared to the results discussed above of highly loaded samples obtained using the shaft/bushing technique. For this test conditions, low friction coefficient at different rotation speed was demonstrated for the samples 4 (PTFE-C) and 5 (PTFE-A) (sample 6 was not tested).

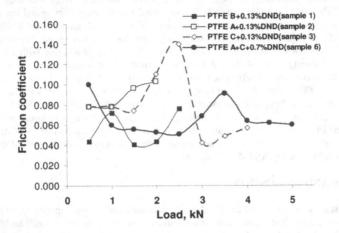

Figure 1. Friction coefficients as a function of a load in the shaft/bushing test for selected samples from Table 1.

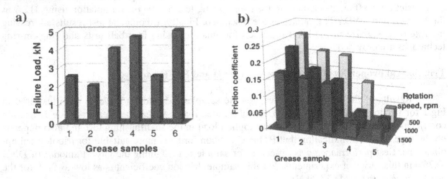

Figure 2. Failure load for the grease samples 1-5 (Table 1) with DND and PTFE additives obtained from shaft/bushing test (a) and friction coefficients for the samples obtained from ring-on-ring test at different rotation speeds and a low load (320N).

Thus, the results from this section demonstrate significantly improved tribological characteristics of Li-grease using DND/PTFE-C (shaft/bushing and ring-on-ring tests) and DND/PTFE-A (in ring-on-ring test) additives at a level of DND load less than 1%. It also demonstrated that depending on the loading/exploitation conditions, lubricant composition, in general, should be fine tuned to provide optimal performance. At the same time, some formulations (like sample 4 with 4% PTFE-C and 0.3%DND) can improve tribological properties of lubricants for different types/level of loading reasonably well.

Tribological properties of the oils with DND and PTFE additives

Results in Figure 3 summarize extremely robust friction-reducing capability of various lubricants using I12A as a base oil with DND at a range of concentrations.

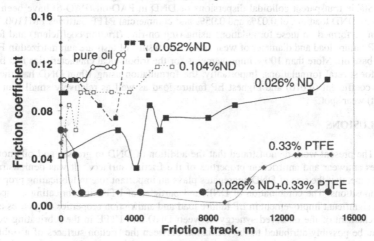

Figure 3. Coefficient of friction of mineral oil (I12A) composition with various amounts of 180nm nanodiamond and PTFE «A» obtained by ring-on-ring technique. Load is 120 N, rpm=1000.

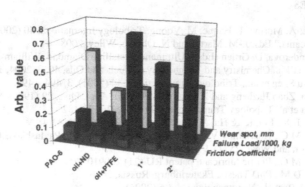

Figure 4. Tribological characteristics of formulations of PAO-6 with DND and 0.3%PTFE (Zonyl MP 1100) additives. Sample 'oil+ND' contains 0.05% DND (150nm); sample 1*: oil+0.3%PTFE+0.03% DND(150nm); sample 2*: oil+0.3%PTFE+0.03% DND(10nm).

As can be seen from Fig.3, the friction coefficient of sliding test pairs in a commercial I12A oil was more than 0.12, but with the addition of DND and PTFE, the friction coefficients were lowered to 0.01 levels when tested in a ring-on-ring test machine. This is an extremely low value. The applied load was 120 N and the rotational velocity was 1000 rpm. Synergistic effect of using ND and PTFE compositions is demonstrated rather clearly in these tests. These tests can be also considered as accelerated life time of the greases. It is clearly demonstrated that the life time of the greases with the nanoadditives was significantly (longer friction track).

Stable transparent colloidal dispersions of DND in PAO oil (PAO-6) have been formulated using DND loadings of 0.03% and 0.05% and commercial PTFE particles PM1100. Testing has been performed on these formulations using ring-on-ring (friction coefficient) and four ball tests (EP failure load and diameter of wear spot). The results of tests are summarized in Fig.4 for PAO-6 base oil. More than 100% improvements for the tribological characteristics had been observed for several formulations. Importantly, the formulation using 10nm DND had the lowest friction coefficient (0.016) and highest EP failure load as well as relatively small (but not the smallest) wear spot.

CONCLUSIONS

The present work demonstrated that the addition of DND to greases and oils noticeably improves antiwear and antifriction properties of the friction surfaces. It was demonstrated for the first time, that the size of DND aggregates plays an important role in lubricating properties of the compositions. The combination of DND and particles of PTFE in lubricating compositions provide significant improvement in EP failure load and antifriction properties of greases and oils. The mechanism of the observed synergy between DND and PTFE in the lubricating compositions can be possibly attributed to the formation between the friction surfaces of a stable polymer-based tribo-film reinforced by DND that is able to withstand multiple deformation modes without failure. Most significant effect of increasing EP failure load was observed for DND in combination with micropowder PTFE subjected to radiation treatment.

REFERENCES

1. Neville, A. Morina, T. Haque, M. Voong, Tribology International 40 (2007) 1680–1695; "Nanolubricants," Eds. J.-M. Martin and N. Ohmae, Wiley, 2008
2. O. Shenderova, D. Gruen (Eds.), Ultrananocrystalline Diamond, William-Andrew (2006)
3. V.E. Red'kin. Chemistry and Technology of Fuels and Oils, 40, 3, 2004, 164 – 179.
4. Mingwu Shen et al... Tribology Trans-actions, 44 (2001), 3, 494 – 498.
5. Xu Tao, Zhao Jiazheng and Xu Kang. J. Phys. D: Appl. Phys. 29 (1996) 2932–2937.
6. Tao Xu et al. Tribology Transactions, 40 (1997), 1, 178-182.
7. Zhang, J. X., Liu, K. & Hu, X. G. (2002). Tribology, 22(1), 44-48.
8. Ivanov M.G., Kharlamov V.V., Buznik V.M., Ivanov D.M., Pavlushko S.G., Tsvetnikov A.K., Friction and Wear, 25 (1), 99 (2004)
9. Ivanov M.G., Plastic lubricant, patent RU (21) 2009105304
10. Ivanov D.M., PhD Thesis, Ekaterinburg, Russia, 2006
11. Puzyr A.P. et al., Nanotechnique, 4, 96 (2006)
12. Buznik V.M.,Rus.Chem.J. (Rus.Chem.J.of the Rus.Chemical Soc. Named by Medeleev), 2008, v. LII, № 3, C.7-12.

Mater. Res. Soc. Symp. Proc. Vol. 1203 © 2010 Materials Research Society   1203-J17-01

Surface potential of functionalised nanodiamond layers

I. Kratochvílová[1], A. Taylor[1], F. Fendrych[1], A. Kovalenko[1] and M. Nesládek[2]

[1]Institute of Physics, Academy of Sciences Czech Republic v.v.i, Na Slovance 2, CZ-182 21, Prague 8, Czech Republic

[2]Hasselt University, Institute for Materials Research (IMO), Wetenschapspark 1, B-3590 Diepenbeek, Belgium

ABSTRACT

Carbon nanomaterials especially ultrananocrystalline diamond and nanocrystalline diamond films have attracted more and more interest due to their unique electrical, optical and mechanical properties, which make them widely used for different applications (e.g. MEMS devices, lateral field emission diodes, biosensors and thermoelectrics). Nanocrystalline diamond can also offer novel advantages for drug delivery development. Recent studies have begun to use nanocrystalline diamond for in-vivo molecular imaging and bio-labeling. To enable grafting of complex bio-molecules (e.g. DNA) the surface of ND requires specific fictionalization (e.g. H, OH, COOH & NH$_2$). Due to the surface dipoles of functionalised nanodiamond band bending at the surface can be easily induced and from the measured band bending we can deduce the type of the fictionalization on the surface. The surface potential of H-terminated and OH terminated nanodiamond layers was investigated by Kelvin probe microscope. From the change of the surface potential value (as the departure of the material surface from the state of electrical neutrality is reflected in the energy band bending) the work function of the H-terminated nanodiamond layer was established to be lower than OH-terminated nanodiamond layer. The surface potential difference can be explained by the surface dipole induced by the electro-negativity difference between the termination atoms.

INTRODUCTION

From carbon nanomaterials specially ultrananocrystalline diamond (UNCD) and nanocrystalline (NCD) diamond films have attracted more and more interest due to their unique electrical, optical, and mechanical properties, which make them widely used for different applications: MEMS devices, lateral field emission diodes, biosensors, thermoelectrics, etc [1-2]. The physico-chemical properties of diamond surfaces are key to achieve the desired properties of these devices. In order to produce high performance devices, therefore, it will be essential to control the surface physico-chemical properties of the diamond.

It has been reported [3] that physico-chemical properties of diamond surfaces are closely related to the surface chemisorbed species on the surface. Hydrogen chemisorption on a chemical vapor deposition -grown diamond surface is well-known to be important for stabilizing diamond surface structures with sp^3 hybridization. Many reports have suggested that an H-chemisorbed structure is necessary to provide a negative electron affinity condition on the diamond surfaces. Oxidized diamond surfaces [4] usually show characteristics completely different from those of the H-chemisorbed diamond surfaces. The unique electron affinity condition, or the surface potential, is thus strongly related to the chemisorbed species on diamond surfaces [3-6]. The relationship between the surface chemisorption structure and the surface electrical properties, such as the surface potential of the diamond, has not been revealed yet.

In this study, changes of the surface potential associated with a change in the chemisorption from hydrogen to oxygen have been observed by Kelvin probe microscopy. ND was first oxidised and after that exposed to an H-plasma. Carbonyl and hydrogen terminated diamond nanoparticles were dropped on gold layer.

KELVIN PROBE

Nanodiamond particles electric surface potential has been measured under room conditions by using special setting of Scanning Probe Microscope - Kelvin Probe Force Microscope (KPFM) NTEGRA Prima NT MDT under ambient conditions .

Kelvin Probe Force Microscope is a hybrid of atomic force microscope (AFM) and the macroscopic Kelvin Method [7, 8] that is, it uses the principles of the Kelvin Method, but with a nanoscale lateral distribution which can be achieved by AFM. In the case of NT-MDT [9,10] system, Kelvin mode is based on the two-pass technique. In the first pass the topography is acquired using standard semicontact mode (mechanical excitation of the cantilever). In the second pass this topography is retraced at a set lift height (10 Å) from the sample surface to detect the electric surface potential $\Phi(x)$. During this second pass the cantilever is no longer excited mechanically but electrically by applying to the tip the voltage V_{tip} containing dc and ac components

$$V_{tip} = V_{dc} + V_{ac} \sin(\omega t)$$

The resulting capacitive force F_{cap} between the tip and a surface at potential V_s is

$$F_{cap} = (1/2)[V_{tip} - \Phi(x)]^2 (dC/dz)$$

where $C(z)$ is the tip-surface capacitance. The first harmonic force

$$F_{cap\,\omega} = (dC/dz)[\ (V_{dc} - \Phi(x))V_{ac} \sin(\omega t)]$$

leads to suitable cantilever oscillations. The feedback then changes the dc tip potential V_{dc} until the ω component of the cantilever (and accordingly ω component of the tip-force) vanishes, e.g. $V_{dc}(x)$ became equal to $\Phi(x)$. So mapping $V_{dc}(x)$ reflects distribution of the surface potential along the sample surface.

We determined the surface potentials of the surface potential of the H-chemisorbed nanodiamonds and the O-chemisorbed nanodiamonds both under adequate oscillation voltage conditions with comparison to the gold layer. The oscillation voltage amplitude was carefully found to prevent errors in the simultaneous topographic tracking observation by excess voltage application. In the KFM observation, the sample bias AC voltage was determined to be 2.5 V_{dc} at

22 kHz. The frequency of the voltage oscillation was slightly lower than the resonant frequency of the cantilever (22.58 kHz) and the forced vibration frequency (22.64 kHz) to obtain high voltage sensitivity. Signals at the voltage oscillation frequency and the vibration frequency are lock-in detected distinctively to obtain a potential and a topographic data.

RESULTS

Diamond nanoparticles for the present experiment (Fig. 1) were prepared by dropping of ND particles on gold layer. ND particles were prepared by detonation synthesis and contain significant amounts of sp^2 carbon. ND was first oxidised to produce carbonyl termination. The ND were then exposed to an H-plasma (2500 W, 600°C, duration 2-5 minutes) in AX6500 SEKI reactor to obtain hydrogen-terminated ND. All the ND termination states were confirmed by ATR-FTIR and XPS.

Figure 1. Schematic illustration of the hydrogen and carbonyl terminated diamond monolayers.

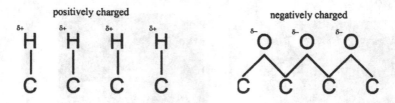

Figure. 2: Schematic representation of the hydrogen and carbonyl terminated diamond surface dipoles.

Figures 3 and 4 show the results of the carbonyl and hydrogen terminated ND Kelvine Probe measurements. The surface potential of hydrogenated nanodiamond particles is bigger (0.1 eV) than that of O-chemisorbed nanodiamonds due to the fact that hydrogen-diamond bonds are polar covalent bonds e.g. surface dipole layer is generated with slightly negative charged carbon

(C-) and slightly positive hydrogen (H+). Such a dipole layer causes an electrostatic potential gap ΔV perpendicular to the surface. As all electronic states are shifted for the same energy, occupied valence band states emerge above the chemical potential of the surrounding. In the case of oxygen-diamond bonds are again polar covalent – dipole layer is made by positively charged carbon and negatively charged oxygen. Occupied valence band are below electronic potential of the surrounding which means that the work function of the –O terminated nanodiamonds surface is larger (surface potential smaller) than the –H terminated nanodiamonds. Here we present surface potential measured on variously (hydrogen and oxygen) terminated nanodiamond particles. We found that the impact of ND surface termination on surface potential is so strong that we can detect the variously terminated ND surface potential change using Kelvin Probe as in the case of nanodiamond layers [3-6]. The surface potential of hydrogenated nanodiamond particles is bigger (0.1 eV) than that of O-chemisorbed nanodiamonds.

A   B

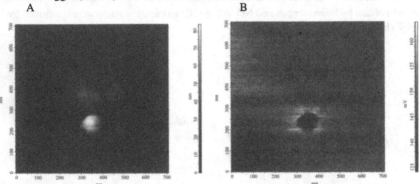

Figure 3. Height (a) and surface potential (b) of oxygen terminated nanodiamond particle on a gold substrate.

A   B

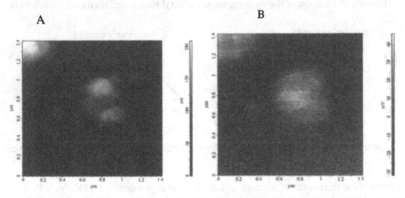

Figure 4. Height (a) and surface potential (b) of hydrogen terminated nanodiamond particle on a gold substrate.

ACKNOWLEDGMENT

This work was supported by the Academy of Sciences of the Czech Republic Grant Nos., KAN401770651, KAN200100801, KAN400480701, the Czech Science Foundation 203/08/1594, the Ministry of Education of the Czech Republic (OC 137), and institutional funds of the Institute of Physics AS CR, v.v.i. AV0Z10100520.

REFERENCES

1. S. Koizumi, Ch. Nebel, M. Nesladek: Physics and Applications of CVD diamond. Wiley-VCH Verlag GmbH&Co KGaA, Weinheim (2008)
2. Ristein J., Nesladek M., Haenen K, Phys. Stat. Sol. A, 204 pp.2835-2835 (2007)
3. Minoru Tachiki, Yu Kaibara, Yu Sumikawa, Masatsugu Shigeno, Hirohumi Kanazawa, Tokishige Banno, Kwang Soup Song, Hitoshi Umezawa, Hiroshi Kawarada, Surface Science vol. 581 pp. 207 - 212 (2005)
4. K. Iakoubovskii, G.J. Adriaenssens, M. Nesladek, J. Phys. C, 12, 189-199 (2000)
5. B. Marczewska, P. Olko, M. Nesladek, MPR Waligorski, Y. Kerremans, Rad. Prot. Dosimetry 101, pp 485-488 (2002)
6. M. Yokoyama, T. Ito, Appl. Surf. Sci. pp. 162 - 163 (2000)
7. H. Gamo, M. N. Gamo, K. Nakagawa and T. Ando: Surface potential change by oxidation of the chemical vapor deposited diamond (001) surface Journal of Physics: Conference Series 61 pp. 327–331 (2007)
8. M. Nonnenmacher, M. P. O'Boyle, and H. K. Wickramasinghe Kelvin probe force microscopy, Appl. Phys. Lett. 25, pp. 2921-2923 (1991)
9. I. Kratochvilova, K. Kral, M. Buncek, A. Viskova, S. Nespurek, A. Kochalska, T. Todorciuc, M. Weiter, B. Schneider, Biophysical Chemistry, vol. 138 pp. 3-10 (2008)
10. M. Kopecek, L. Bacakova ,J. Vacik, F. Fendrych,V. Vorlicek,I. Kratochvilova, V.Lisa, E. Van Hove , C. Mer, P.Bergonzo, M. Nesladek, Phys. Stat. Sol. A, 205, 2146-2153 (2008)

Mater. Res. Soc. Symp. Proc. Vol. 1203 © 2010 Materials Research Society   1203-J17-52

Purification and Functionalization of Diamond Nanopowders

Jong-Kwan Lim and Jong-Beom Baek
School of Energy Engineering, Ulsan National Institute of Science and Technology (UNIST), 100, Banyeon, Ulsan, 689-805 South Korea

ABSTRACT

Purification of diamond nanopowder (DNP) was conducted in a less-destructive mild polyphosphoric acid (PPA)/phosphorous pentoxide (P_2O_5). The wide-angle X-ray diffraction (XRD) showed that the intensity of the characteristic diamond d-spacing (111) at 2.07 Å from purified DNP (PDNP) was fairly increased compared to pristine DNP, indicating that significant amount of carbonaceous impurities were removed. Chemical modification of pristine DNP and PDNP with 4-ethylbenzoic acid was carried out to afford 4-ethylbenzoyl-functionalized DNP (EBA-g-DNP) and PDNP (EBA-g-PDNP). The morphologies of EBA-g-DNP and EBA-g-PDNP from scanning electron microscopy (SEM) were further affirmed the feasibility of chemical modification. The results suggested that the reaction condition was indeed viable for the one-pot purification and functionalization of DNP. The resultant functionalized DNP could be useful for nanoscale additives. Hence, EBA-g-DNP and EBA-g-PDNP was brominated by using N-bromosuccinimide (NBS). The resultant α-brominated DNP and PDNP could be used as initiator for the atom transfer radical polymerization (ATRP) to introduce many polymers onto the surface of functionalized DNP and PDNP.

1. INTRODUCTION

Carbon nanomaterials have received a great deal of attention since the discovery of the C_{60} fullerene molecule [1] and the carbon nanotube [2]. Three-dimensional, carbon-based, nano-structured materials are generally various: (i) fullerene (0.7~3 nm); (ii) carbon nanopowders (≤30 nm); (iii) diamond nanopowders (3.2 nm). Amongst them, diamond nanopowders (DNP) are worth of investigating due to relatively new and thus less studied. DNP has unique properties such as highest hardness, highest thermal conductivity, low thermal expansion coefficient and low friction coefficient. DNP is generally produced by chemical vapor deposition (CVD), detonation and spark plasma sintering method [3]. Metal catalysts are generally necessary to activate materials (DNP: ≥95% trace metals basis). In consequence, DNP from those methods contains impurities such as metal catalyst particles, amorphous carbon and carbon nanoparticles. Therefore, DNP may have limited applications because it contains impurities. Lately, many purification methods have been reported to remove impurities from carbon soot [4,5]. For example, they are oxidation, thermal annealing, thermal oxidation, ultra-sonication, strong acid treatments and so on. Recently, less-destructive method for both purification and chemical modification of various carbon nanomaterials in a mild polyphosphoric acid (PPA)/phosphorous pentoxide (P_2O_5) medium have been developed [6].

Hence, as-received DNP was treated in PPA/ P_2O_5 to selectively remove persisting impurities. In the same purification condition, DNP and purified DNP (PDNP) were functionalized as followed by previously reported procedure [7,8]. Therefore, pristine DNP and purified DNP (PDNP) were treated with 4-ethylbenzoic acid (EBA) to afford 4-ethylbenzoyl-functionalized DNP (EBA-DNP) and PDNP (EBA-PDNP). Furthermore, EBA-DNP and EBA-PDNP were treated with *N*-bromosuccinimide (NBS) to prepare α-brominated DNP and PDNP, which could be used as initiator to graft many polymers onto the surface of DNP and PDNP via atom transfer radical polymerization (ATRP).

2. EXPERIMENTAL DETAILS

2-1. Purification of DNPs in PPA/P$_2$O$_5$

Into a 100mL resin flask equipped with a high torque mechanical stirrer, nitrogen inlet and outlet, DNPs (1g), PPA (83% P$_2$O$_5$ assay; 10g), and phosphorus pentoxide (P$_2$O$_5$, 5g) were placed and stirred under dry nitrogen purge at 130 °C for 72h. After cooling down to room temperature, water was added to the reaction mixture. The resulting precipitate was collected and Soxhlet extracted with water for three days, and then with methanol for three days. Finally, the PPA-treated DNPs (PDNP) were freeze-dried under reduced pressure (0.5mm Hg) for 24h to afford 0.92 g (92% yield) of a black powder.

2-2. Functionalization of DNPs with 4-ethylbenzoic acid in PPA/P$_2$O$_5$ at 130 °C (EBA-g-DNP)

Into a 100mL resin flask equipped with a high torque mechanical stirrer, nitrogen inlet and outlet, EBA (0.5g), DNP (0.5g), PPA (83% P$_2$O$_5$ assay; 20g), and phosphorus pentoxide (P$_2$O$_5$, 5g) were placed and stirred under dry nitrogen purge at 130 °C for 72h. After cooling down to room temperature, water was added to the reaction mixture. The resulting precipitate was collected and Soxhlet extracted with water for three days, and then with methanol for three days. Finally, the resultant powder was freeze-dried under reduced pressure (0.5mm Hg) for 24h to afford to 0.63 g (67 % yield) of a dark gray powder.

2-3. Functionalization of PDNP with 4-ethylbenzoic acid in PPA/P$_2$O$_5$ at 130 °C (EBA-g-PDNP)

The modification procedures are the same as those described in Section 2-2 except that PDNP was used instead of DNP.

2-4. Bromination of EBA-g-DNP using NBS (EBA-g-DNP (Br))

Into a 50mL three neck round-bottom flask equipped with a magnetic bar, nitrogen inlet and outlet, EBA-g-DNP (0.5g), NBS (1.4g), BPO (0.13g), and CCl$_4$ (8ml) were heated under reflux for 1h. The reaction mixture was then cooled in an ice bath and the resulting precipitate was collected and Soxhlet extracted with water for four days, and then with methanol for one days. Finally, the resultant powder was freeze-dried under reduced pressure (0.5mm Hg) for 24h to afford to 0.5 g of a dark brown powder.

2-5. Bromination of EBA-g-PDNP using NBS (EBA-g-PDNP (Br))

The bromination procedures are the same as those described in Section 2-4 except that EBA-g-PDNP was used instead of EBA-g-DNP.

2-6. PPA treated EBA

The purification procedures are the same as those described in Section 2-1 except that no DNP was involved.

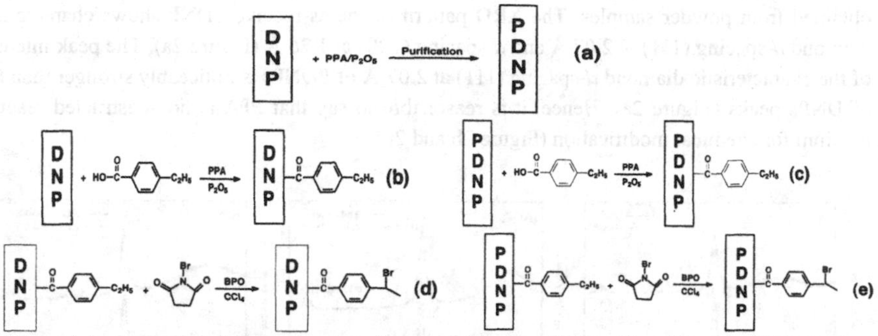

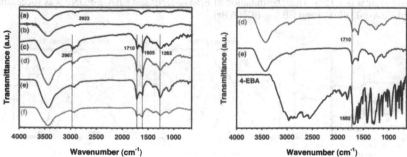

Scheme 1. (a) purification of diamond nanopowder; (b) modification of DNP ; (c) modification of PDNP ; (d) bromination of EBA-g-DNP ; (e) bromination of EBA-g-PDNP

3. RESULTS AND DISCUSSION

FT-IR

Figure 1 shows the FT-IR spectra of samples. The FT-IR spectra of pristine DNP and PDNP (Figure 1a and 1b) show the weak sp^2C-H and sp^3C-H bands around 2923 cm^{-1}. The sites are for Friedel-Crafts acylation reaction. After grafting of EBA, new C=O (carbonyl) peak appears at 1710 cm^{-1} (Figure 1c and 1d), aromatic C=C peak appears at 1605 cm^{-1} and C=O peak in 4-EBA shifted as shown in the right. On the basis of FT-IR result, the functionalization of DNPs and PDNP with EBA was confirmed. It is difficult to detect Br peaks from brominated samples (Figure 1e and 1f), since bromine stretching is weak.

Figure 1. FT-IR spectra: (a) DNP; (b) PDNP; (c) EBA-g-DNP; (d) EBA-g-PDNP; (e) EBA-g-DNP(Br); (f) EBA-g-PDNP(Br).

XRD

To monitor crystallinity and purity after PPA treatment, XRD scattering patterns were obtained from powder samples. The XRD pattern of the as-received DNP shows characteristic diamond *d*-spacing (111) at 2.07 Å and *d*-spacing (220) at 1.26 Å (Figure 2a). The peak intensity of the characteristic diamond *d*-spacing (111) at 2.07 Å of PDNPs is noticeably stronger than that of DNPs peaks (Figure 2a). Hence, it is reasonable to say that PPA is non-destructed reaction medium for chemical modification (Figure 2b and 2c).

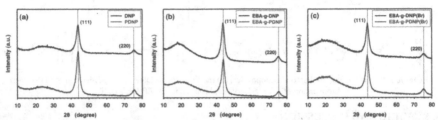

Figure 2. X-ray diffraction patterns of samples

XPS

The XPS spectra of the pristine DNP and functionalized DNP were also measured (Figure 3a). After purification, the peak centered at 287.8eV from DNPs is significantly different from the peak at 287.4eV of PDNPs. It is probably due to surface oxidation by PPA. And we can confirm that pristine DNP have impurities such as C-O, C=O, C-H....etc. (Figure 3b). After modification, the peak position of EBA-g-DNP was shifted at 284.9eV. Figure 3c shows that EBA-g-DNP was fitted with diamond C, graphite C, C=O (combined C_6H_5). Therefore, we can say pristine DNP was successfully functionalized to EBA-g-DNP due to the C=O (combined C_6H_5). The Br3d at 70.5eV (Br; combined carbon) was observed after bromination. The quantity of measured Br was 1.15%. Thus, bromination of EBA-g-DNP and EBA-g-PDNP was successful to afford EBA-g-DNP(Br) and EBA-g-PDNP(Br).

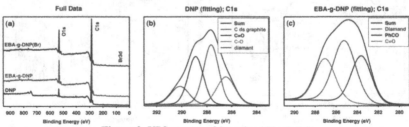

Figure 3. XPS spectra of Samples (C1s, Br3d)

SEM

The SEM images of pristine DNP shows that the surface texture consisted of aggregates and agglutinates (Figure 4a). After purification in PPA, the surface texture of PDNP displayed more smooth and uniform than pristine DNP (Figure 4b). The functionalized samples, EBA-g-DNP and EBA-g-PDNP, show that particles are less aggregates and agglutinates than pristine DNP and PDNP (Figure 4c and 4d). Due to covalent attachment of EBA on the surface, dispersibility of EBA-g-DNP and EBA-g-PDNP should be improved. Brominated samples EBA-g-DNP(Br) and EBA-g-PDNP(Br) show slightly rougher surfaces with similar particle size (Figure 4e and 4f).

Figure 4. SEM images of (a) DNP; (b) PDNP; (c) EBA-g-DNP; (d) EBA-g-PDNP; (e) EBA-g-DNP(Br); (f) EBA-g-PDNP(Br). Scale bars are: 500nm)

TGA

The samples were subject to thermogravimetric analysis (TGA) with heating rate of 10 °C/min in air. The PDNPs are thermally more stable than pristine DNPs above onset temperature of oxidative decomposition. It is probably due to the removal of some thermooxidatively unstable impurities from the pristine DNPs during PPA treatment. The DNPs, PDNP, EBA-g-DNP, EBA-g-PDNP, EBA-g-DNP(Br) and EBA-g-PDNP(Br) showed that the temperature at which 5% weight loss ($T_{d5\%}$) in air was occurred at 578, 562, 445, 426, 252 and 298 °C, respectively (Figure 5).

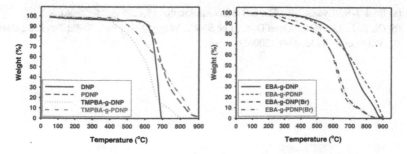

Figure 5. TGA thermograms obtained with heating rate of 10 °C/min in air.

4. CONCLUSIONS

Purification and functionalization of pristine DNPs was conducted at same reaction condition to afford purified PDNP and functionalized DNPs (EBA-g-DNP and EBA-g-PDNP). The degree of functionalization and structure of functionalized DNPs were confirmed by thermal, spectroscopic and microscopic analyses. Hence, the reaction condition should be viable for one-pot purification and functionalization of DNP for verification application. Thus, it has been demonstrated that EBA-g-DNP and EBA-g-PDNP can be brominated to EBA-g-DNP(Br) and EBA-g-PDNP(Br), which are useful initiator for atom transfer radical polymerization (ATRP).

5. ACKNOWLEDGMENTS

The Korea Science and Engineering Foundation (KOSEF) (2009-0052351)
Korea Research Foundation (KRF) (D00267)
Ulsan National Institute of Science and Technology (UNIST)
World Class University (WCU) Program

6. REFERENCES

1. H. W. Kroto.; J. R. Heath.; S. C. O'Brien.; R. F. Curl.; and R. E. Smally. *Nature* (London) 318, 162 (1985).
2. S. Iijima. *Nature* (London) 354, 56 (1974).
3. Faming Zhang.; Jun Shen.; Jianfei Sun.; Yan Qiu Zhu.; Gang Wang.; G. McCartney. *Carbon* 43, 1254-1258 (2004).
4. R. Andrews.; D. Jacques.; D. Qian.; E.C. Dickey. *Carbon* 39, 1681 (2001).
5 . P.X. Hou.;   S. Bai.;   Q.H. Yang.;   C. Lin.;   H.M. Cheng. *Carbon* 40, 81 (2002).
6 . Baek, J.-B.; Lyon, C. B.; Tan, L.-S. *J. Mater. Chem.* 14, 2052 (2004).
7. Ahn, S.-N.; Lee, H.-J.; Kim, B.-J.; Tan, L.-S.; Baek, J.-B. *J. Polym. Sci. Pol. Chem.* 46,7473–7482 (2009).
8 . (a) Baek J-B.; Lyon C.-B.; Tan L.S. Macromolecules 37,8278-8285 (2006).
    (b) Oh S.-J.; Lee H.-J.; Keum D.-K., Lee S.-W.; Wang D.-H.; Park S.-Y.; Tan L.S.; Baek J.-B. . Polymer 47,1132–1140 (2006).

Graphene

Mater. Res. Soc. Symp. Proc. Vol. 1203 © 2010 Materials Research Society   1203-J09-02

Nanoscale modification of graphene transport properties by ion irradiation

F. Giannazzo [1], S. Sonde[1,2],V. Raineri[1], E. Rimini[1,3]
[1]CNR-IMM, Strada VIII, 5, 95121, Catania (Italy)
[2]Scuola Superiore di Catania, Via San Nullo, 5/i, 95123, Catania (Italy)
[3]Dipartimento di Fisica ed Astronomia, Università di Catania, Via S. Sofia, 64, 95123, Catania (Italy)

ABSTRACT

Single layers of graphene (SLG) mechanically exfoliated from highly oriented pyrolytic graphite and deposited on SiO_2/Si were irradiated with C^+ ions at different fluences (from 10^{13} to 10^{14} cm^{-2}), in order to modify the transport properties in controlled way. Using a method based on scanning probe microscopy, local measurements of the electron mean free path (l) have been carried out both on pristine and ion irradiated SLG. A lateral inhomogeneity of l was found in both cases, with an increasing spread in the distribution of l for larger fluences. Before irradiation, the spread was explained by the inhomogeneous distribution of charged impurities on SLG surface and/or at the interface with SiO_2. After irradiation, lattice vacancies cause a local reduction of l in the damaged regions.

INTRODUCTION

Graphene is the subject of great research interest, due to its outstanding transport properties making it an attractive candidate for post-Si electronics [1-4]. In the case of ideal (i.e., free-standing, defects and impurities free) graphene layers, values of the electron mobility $\mu \approx 2 \times 10^5$ $cm^2V^{-1}s^{-1}$ (corresponding to an electron mean free path $l \approx 2.3$ μm) have been estimated for a carrier density $n \approx 10^{12}$ cm^{-2} and at room temperature (300 K) [3]. However, in most practical cases, graphene is placed on a substrate (SiO_2 or other dielectrics) and is subjected to lithographic processes (for patterning and/or contact formation), which leave chemical residuals. In these cases, the reported μ values are typically ranging from 10^3 to 10^4 $cm^2V^{-1}s^{-1}$, corresponding to mean free paths from ~10 to ~100 nm. Many experimental [3] and theoretical [5,6] works are focused on the role of the different scattering mechanisms limiting l and μ (i.e. scattering by charged impurities, point defects, lattice phonons…). The screened Coulomb interaction between charged impurities and the two-dimensional-electron-gas (2DEG) in graphene causes "long range" scattering and represents the main mechanism limiting l in most of practical cases. However, there is a significant interest on the role played by point defects, which are sources of "short range" scattering [7]. In this context, plasma treatments [8], electron [9] or ion irradiation [10,11] have been proposed as means to introduce defects in single layers of graphene. Among all of these methods, ion irradiation allows a better control of the damage, through the control of the irradiated fluence.

In this paper, irradiation with energetic ions was carried out on single layers of graphene (SLG) on SiO_2 and the effect on the electron mean free path was investigated as a function of ion fluence. While μ and l are commonly obtained by measurements on micrometer size test patterns fabricated on SLG (and represent "average" values on the pattern area), here l was probed "locally" on sub-micrometer scale by scanning probe microscopy [12]. Measurements were carried out on several points both on pristine samples and on irradiated ones at different fluences, and the histograms of the l values were compared. Before irradiation, the spread in the distribution of l was explained by the local changes in the density of charged impurities on SLG surface and/or at the interface with SiO_2. After irradiation, a larger spread in the l distribution was obtained, with a broad tail at low values, which was ascribed to defects (mainly lattice vacancies) causing a local reduction of l.

EXPERIMENT

Graphene samples obtained by mechanical exfoliation of HOPG were deposited on a n^+-Si substrate covered with 100 nm thermally grown SiO_2 [13]. Optical microscopy, tapping mode atomic force microscopy (AFM) and micro-Raman spectroscopy were used to identify SLG. As-deposited samples were irradiated with C^+ ions at 500 keV. Irradiations were carried out under high vacuum conditions (10^{-6} Torr) in order to minimize surface contaminations. At 500 keV energy the projected range of the ions is ~1 μm, quite deep into the n^+-Si substrate. This minimizes the damage both in the 100 nm SiO_2 layer and at the interface between SiO_2 and n^+ Si. Different ion fluences, ranging from 1×10^{13} to 1×10^{14} ions/cm^2, were used. The energy released to graphene by incident C^+ ions is shared between two different contributions: (i) collisions with the C atoms of graphene lattice (elastic energy loss), and (ii) interaction with electrons (inelastic energy loss). For 500 keV C^+ ions, only a small fraction of the ion energy is lost by direct C^+-C collisions (~0.6eV), whereas most of the energy is lost by the inelastic collisions with electrons in graphene (~270 eV). This energy is in turn converted into lattice vibrations, i.e. a "local heating" of graphene occurs. The latter can promote local changes in the adhesion of graphene with the SiO_2 substrate.

After irradiation, the samples were characterized with high-resolution AFM measurements in order to investigate the morphological modifications induced by the ion beam both on the SLG sheets and on the SiO_2 surface. No significant variations in the roughness of bare SiO_2 was found after irradiation. On the contrary, on SLG a significant variation in the roughness was observed from pristine to irradiated samples. This increase in the roughness can be mainly explained in terms of the change of the adhesion of graphene on the substrate. Micro Raman spectroscopy was used also to estimate the amount of disorder in the sheets, from the intensity of the characteristic D peak at 1350 cm^{-1}, commonly associated to defects in graphene.

The ion irradiation induced changes in graphene electronic properties were investigated by local capacitance measurements on the SLG/SiO_2/n^+Si stack, using scanning capacitance spectroscopy (SCS) [12-14]. SCS was performed at room temperature using a DI3100 AFM by Veeco equipped with Nanoscope V electronics and with the scanning capacitance microscopy (SCM) head. A conductive AFM tip (Pt coated Si tips were used in this experiment) is placed on a discrete array of positions, lifting the tip by 20 nm at every interval. This 'step and measure' approach eliminates the lateral (shear) force usually present when tip is scanned on a surface. Moreover, the vertical contact force can be suitably minimized to get a good electrical contact to the SLG while avoiding damage at the same time. A modulating bias $\Delta V=V_g/2[1+\sin(\omega t)]$, with amplitude V_g in the range from 0 to 1.2 V and frequency $\omega=100$ kHz, was applied between the Si n^+ backgate and the nanometric tip contact on SLG (see the schematic in Fig.1(a)). An ultra-high-sensitive (10^{-21} F/Hz$^{1/2}$) capacitance sensor connected to the tip measures, through a lock-in system, the capacitance variation ΔC induced by the modulating bias for several positions of the tip.

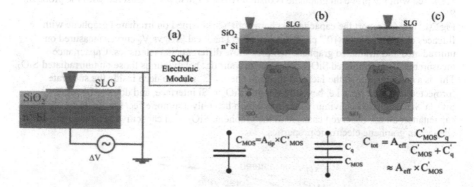

Fig.1 Schematic description of the SCS experimental set-up

When the probe is in contact with bare SiO_2, the tip-SiO_2-semiconductor system can be described as a nanoscale Metal-Oxide-Semiconductor (MOS) capacitor (see Fig.1(b)) with capacitance $C_{MOS}=A_{tip}C'_{MOS}$, being A_{tip} the tip "geometrical" contact area ($A_{tip}\approx 80$ nm^2) and C'_{MOS} the MOS capacitance per unit area.

The graphene/SiO_2/Si stack can be described as a capacitor with effective area A_{eff} and with capacitance per unit area given by the series combination of C'_{MOS} and C'_q (see Fig.1(c)). The latter contribution is the quantum capacitance per unit area associated to the graphene 2DEG [14]. It has been shown that $C'_q>>C'_{MOS}$, for a SiO_2 thickness $t_{ox}=100$ nm [14]. Hence, the total capacitance of the graphene/SiO_2/Si capacitor is $C_{tot}\approx A_{eff}C'_{MOS}$, i.e. the presence of graphene between the tip and SiO_2 leads to an increase of the capacitance by a factor given by A_{eff}/A_{tip}. As discussed in Ref. [14], the AC component of the applied bias induces an excess of electrons in the graphene region under the tip; these electrons spread (diffuse) in the region around the tip over the area A_{eff}. The values of A_{eff} at each tip position on graphene can be obtained from the ratio of the local $|\Delta C_{tot}|$-V_g curves with the average $|\Delta C_{MOS}|$-V_g curve [14] measured with the tip on the SiO_2, i.e. $A_{eff}\approx A_{tip}|\Delta C_{tot}|/|\Delta C_{MOS}|$. It is worth noting that A_{eff} is typically larger than the tip contact area, but much smaller than the total graphene flake area. Assuming the effective area of circular shape, an effective length $L_{eff}=(A_{eff}/\pi)^{1/2}$ can be defined. It has been shown recently that, for large enough gate bias (i.e. far from the Dirac point), L_{eff} corresponds to the local electron mean free path l in graphene [12]. This value of l has to be considered as the mean free distance traveled by an electron between "few" scattering events.

RESULTS AND DISCUSSION

In the following the capacitive characteristics measured when the tip is on bare SiO_2 and on graphene covered regions are reported. In Fig.2(a), 25 capacitance characteristics measured on an array of 5×5 positions on an unirradiated sample with the inter-step interval of 1μm×1μm are reported. A capacitance curve on bare SiO_2 is reported for comparison. It is evident that $|\Delta C|$

measured with the probe on graphene is distinctly higher than $|\Delta C|$ measured with the probe on SiO_2.

Fig.2(b) and (c) report the capacitance characteristics obtained on irradiated graphene with fluences of 1×10^{13} and 1×10^{14} cm^{-2}, respectively. The local $|\Delta C|$ vs V_g curves measured on unirradiated and irradiated graphene samples show different characteristics. Capacitance measurements on irradiated SiO_2 show instead a similar behaviour as those on unirradiated SiO_2. This is a consequence of the fact that the high energy C^+ ions go deep inside the substrate (projected range ≈ 1 μm), i.e. beyond SiO_2 and SiO_2/n^+Si interface, and do not affect the SiO_2/n^+Si capacitive behaviour. Since irradiation has only a minor effect on the SiO_2/n^+Si MOS capacitance, all the observed changes in the graphene/SiO_2/n^+Si capacitive behaviour are due to changes in graphene electronic properties.

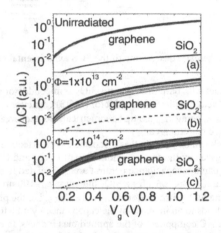

Fig.2 $|\Delta C|$ vs V_g measurements on arrays of several tip positions on unirradiated (a) and on irradiated graphene at fluences 1×10^{13} (b), and 1×10^{14}cm^{-2} (c) respectively. Curves measured with the tip on SiO_2 reported for comparison.

Indeed, the SCS curves on irradiated graphene exhibit noticeably different behavior with respect to those on unirradiated graphene, showing widely varying signal at different points on the sheet. This varying capacitive behavior is an early indication of the changes in the local electronic properties in graphene due to the irradiation induced local damage.

In the following the local mean free paths extracted from the measured $|\Delta C|$-V_g characteristics, according to the procedure illustrated in Ref. [12], will be considered.

In Fig.3(a) we report the l vs. V_g curves for the unirradiated sample, whereas Fig.3(b) and (c) show the l-V_g curves for the samples irradiated with the lowest and highest fluences. l is plotted also as a function of the 2DEG density n, that is related to V_g as $n=C'_{ox}V_g/q$, being $C'_{ox}=\varepsilon_0\varepsilon_{ox}/t_{ox}$, with ε_0 the vacuum dielectric constant, ε_{ox} the relative dielectric constant and t_{ox} the thickness of the SiO_2 layer.

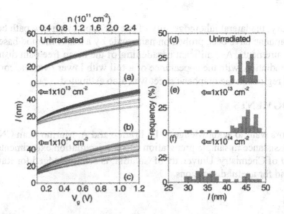

Fig. 3 Local electron mean free path l versus V_g or the carrier density n for different tip positions on unirradiated (a) and irradiated graphene with fluences 1×10^{13} cm^{-2} (b) and 1×10^{14} cm^{-2} (c). Histograms of l at fixed V_g (1V) for unirradiated (d) and irradiated samples with the two fluences (e), (f).

Recently, the dependence on the carrier density of the electron mean free path in graphene has been discussed in details, in the framework of a semiclassical model based on the Boltzmann transport theory [5,6], considering the main scattering mechanism limiting the electron mean free path in graphene, i.e. Coulomb scattering by charged impurities, scattering by phonons and scattering by point defects. While the electron mean free path limited by charged impurities and vacancies scattering was expected to increase as $n^{1/2}$, the mean free path limited by scattering with phonons is expected to decrease as $1/n$. The obtained results show an increase of l with $n^{1/2}$, indicating that the main scattering mechanisms are represented by the scattering with charged impurities and/or with vacancies [12]. In unirradiated graphene, the vacancy contribution can be neglected. The main mechanism limiting l is represented by the charged impurities scattering. These impurities are either adsorbed on graphene surface or located at the interface with SiO$_2$ substrate. Their random spatial distribution is responsible of the observed spread in the local values of l. On irradiated graphene, scattering with point defects plays a significant role and it is responsible of reduction of l for some probed positions. It is worth noting that the number of probed positions with reduced l increases with the fluence.

Finally, in Fig.3 (d), (e) and (f), the histograms of the l (at fixed bias V_g=1V) on unirradiated and irradiated graphene with the lowest and highest fluence are reported. For unirradiated samples the distribution exhibits a single peak at $l_0 \approx 45$ nm with FWHM$\approx$4 nm. In irradiated graphene, in addition to a narrow peak at l_0 (associated to the defect-free sample regions), a broader distribution extending from 27~ to ~40 nm and centered around 34 nm is obtained. The number of counts under the peak at l_0 decreases with increasing the irradiation fluence, whereas that under the broad distribution increases.

CONCLUSIONS

In summary, the lateral inhomogeneity of the electron mean free path both in pristine and ion irradiated graphene have been probed on nanoscale by a novel method based on local capacitance measurements. A significant broadening of the mean free path distribution is observed after irradiation, with the appearance of a tail with lower l values, corresponding to the locally damaged regions in the graphene sheet was also estimated.

ACKNOWLEDGMENTS

The authors want to acknowledge S. Di Franco and A. Marino from CNR-IMM, Catania, for their expert assistance in sample preparation and ion irradiation experiments. G. Compagnini from Department of Chemistry, University of Catania, is acknowledged for Raman measurements and for useful discussions.

REFERENCES

1. K. S. Novoselov, A. K. Geim, S. V. Morozov, D. Jiang, Y. Zhang, S. V. Dubonos, I. V. Grigorieva, A. A. Firsov, *Science* 306, 666 (2004).
2. Y. Zhang, Y.-W. Tan, H. L. Stormer, P. Kim, *Nature* 438, 201 (2005).
3. J.H. Chen, C. Jang, S. Xiao, M. Ishigami, M. S. Fuhrer, *Nature Nanotechnol*, 3, 206 (2008).
4. Z. Jiang, E. A. Henriksen, L. C. Tung, Y.-J. Wang, M. E. Schwartz, M. Y. Han, P. Kim, H. L. Stormer, *Phys. Rev. Lett.*, 98, 197403 (2007).
5. T. Stauber, N. M. R. Peres, F. Guinea, *Phys. Rev.* B, 76, 205423 (2007).
6. E. H. Hwang, S. Adam, and S. Das Sarma, *Phys. Rev. Lett.* 98, 186806 (2007).
7. V. M. Pereira, J. M. B. Lopes dos Santos, A. H. Castro Neto, *Phys. Rev. B* 77, 115109 (2008).
8. K. Kim, H. J. Park, B.-C. Woo, K. J. Kim, G. T. Kim, W. S. Yun, *Nano Lett.* 8, 3092 (2008).
9. D. Teweldebrhan, A. A. Balandin, *Appl. Phys. Lett.* 94, 013101 (2009).
10. L. Tapasztó, G. Dobrik, P. Nemes-Incze, G. Vertesy, Ph. Lambin, L. P. Biró, *Phys. Rev. B*, 78, 233407 (2008).
11. J.-H. Chen, W. G. Cullen, C. Jang, M. S. Fuhrer, E. D. Williams, *Phys. Rev. Lett.*, 102, 236805 (2009).
12. F. Giannazzo, S. Sonde, V. Raineri, and E. Rimini, *Appl. Phys. Lett.* 95, 263109 (2009).
13. S. Sonde, F. Giannazzo, V. Raineri, E. Rimini, *J. Vac. Sci. Technol. B*, 27, 868-873 (2009).
14. F. Giannazzo, S. Sonde, V. Raineri, and E. Rimini, *Nano Lett.* 9, 23 (2009).

Doping

Mater. Res. Soc. Symp. Proc. Vol. 1203 © 2010 Materials Research Society   1203-J17-17

Single Crystal Boron-Doped Diamond Synthesis

T. A. Grotjohn[1], S. Nicley[1], D. Tran[1], D. K. Reinhard[1], M. Becker[2] and J. Asmussen[1,2]

[1]Michigan State University, Electrical and Computer Eng., East Lansing, MI 48824, U.S.A.
[2]Fraunhofer USA Center for Coatings and Laser Applications, East Lansing, MI 48824, U.S.A.

ABSTRACT

The electrical characteristics of high quality single crystal boron-doped diamond are studied. Samples are synthesized in a high power-density microwave plasma-assisted chemical vapor deposition (CVD) reactor at pressures of 130-160 Torr. The boron-doped diamond films are grown using diborane in the feedgas at concentrations of 1 to 50 ppm. The boron acceptor concentration is investigated using infrared absorption and a four point probe is used to study the conductivity. The temperature dependent conductivity is analyzed to determine the boron dopant activation energy.

INTRODUCTION

Diamond's exceptional properties, such as a wide bandgap, high breakdown voltage, and high electron and hole mobilities, make it a potentially useful semiconductor for high-temperature and high-power devices. The realization of useful devices requires the deposition of high quality, controlled conductivity films. Our previous work [1,2] on the deposition of high-quality boron-doped single crystal diamond measured the growth rates of boron-doped films as a function of the concentration of diborane and methane in the feedgas, with the aim of increasing the growth rate and the quality of the films. This earlier work demonstrated the deposition of thick boron doped layers up to 2 mm thick at rates up to 11.5 μm/hr. The deposition pressure used in this previous work was 135-160 Torr, which is higher than many other studies of boron doped diamond, as described in [1,2]. This work expands upon our previous effort by performing more electrical characterization of plasma-assisted CVD deposited diamond.

EXPERIMENT

Boron doped diamond was deposited on 3.5 mm x 3.5 mm HPHT diamond seeds using a microwave plasma-assisted CVD reactor [3]. The reactor operates at 2.45 GHz with a molybdenum substrate holder that is water cooled. The deposition conditions included pressures of 130-160 Torr, 1-50 ppm diborane concentration in the methane/hydrogen feedgas, 4-7.5% methane, and substrate temperatures of 800-1100 C. The seed substrates were acid cleaned and hydrogen plasma etched before deposition.

Infrared Absorption

The deposited diamond films were characterized using FTIR measurements to quantify the boron-related infrared (IR) absorption. IR absorption spectroscopy has proven to be a simple, nondestructive method of probing the electronic structure of semiconducting diamonds [4-7].

Collins and Williams [4] were the first to use the area under the absorption band at 347 meV (corresponding to transitions from the ground state to the excited states of the bound hole) to determine the concentration of boron in diamond. Using Hall Effect data taken on natural type IIb diamond, a calibration curve was developed. Interpretation of the figures in this reference gives a linear relationship between the integrated absorption (the area integrated under the 347 meV (= 2802 cm^{-1}) absorption peak), and the uncompensated acceptor concentration given in equation 1

$$N_A - N_D \ (units:cm^{-3}) = 7.82 \times 10^{17} I (347 \, meV)(units:eV \cdot cm^{-1}) \qquad (1)$$

where I(347 meV) is the integrated absorption of the peak at 347 meV in units of eV·cm^{-1}. For higher boron concentrations and/or thick samples, the peak at 347 meV becomes more difficult to measure because the infrared transmission drops toward 0% as the incorporated boron increases. To determine the concentration in these samples, Blank et. al [7] used a relationship between the uncompensated boron concentration and the height of the peak at 160 meV (=1290 cm^{-1}), as given in equation 2.

$$N_A - N_D \ (units:cm^{-3}) \approx 1.6 \times 10^{17} \alpha \, (1290 cm^{-1}) (units:cm^{-1}) \qquad (2)$$

A later study by Gheeraert et al. [5] found that the boron concentration (at higher boron concentrations than those studied by Collins and Williams) is related by a different relationship for the area under the same peak, given in equation 3.

$$N_A - N_D \ (units:cm^{-3}) = 1.1 \times 10^{15} I (347 \, meV)(units:cm^{-2})$$
$$= 8.9 \times 10^{18} I (347 \, meV)(units:eV \cdot cm^{-1}) \qquad (3)$$

This relationship from Gheeraert et.al. was determined based on a comparison of their IR data with SIMS (Secondary Ion Mass Spectroscopy) data. A later review by Thonke [6] examined this calibration, and according to Thonke, SIMS might overestimate the concentration of electrically active, uncompensated, substitutional boron. For the samples evaluated in the present study, the relationships of equations 2 and 3 were utilized. An example FTIR spectrum and the associated absorption signal are shown in Fig. 1.

Four Point Probe

The conductivity of the deposited diamond was determined from the sheet resistivity measured with a four point probe [8,9] and the thickness of the grown films. The correction factor for the geometry of the sample was determined based on the work of Smits [8]. The conductivity is a function of temperature, and the temperature dependant conductivity of a diamond sample gives additional information about the boron dopants, including the activation energy. Lagrange et.al. [10] gives, for data taken at low temperatures (such that the hole concentration, p, is less than the compensation density, N_D, i.e., $p << N_D$), that the low temperature conductivity (σ_1) is related to the activation energy (E_A) by the relationship of equation 4.

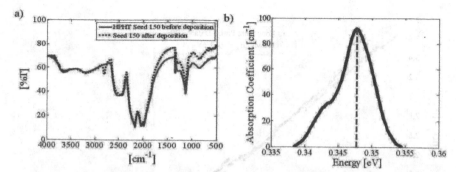

Fig. 1: a) The FTIR spectra of sample 150 before and after boron-doped diamond deposition. A substantial increase in absorbance is noted at the 2802 cm^{-1} peak in the sample after deposition, corresponding to the 347 meV absorption peak associated with the boron acceptor. b) The corresponding absorption peak for sample 150 at 347 meV, as used to calculate the integrated absorption.

$$\sigma_1 = q\mu \frac{N_V}{g_a}\left(\frac{N_A - N_D}{N_D}\right)\exp\left(-\frac{E_A}{kT}\right)$$

$$\text{with}\begin{cases}\mu = \mu_{300}\left(\dfrac{T}{300}\right)^{-s} \\[2mm] N_V = N_0 T^{3/2}\end{cases} \qquad\qquad (4)$$

where q is the electronic charge, N_A and N_D are the acceptor and compensating donor concentrations, respectively, T is the temperature, N_V is the density of states of the valence band, g_a is the degeneracy factor for the acceptor, μ is the conductivity mobility, μ_{300} is the mobility at 300 K, s is a constant (fitted to 2.7 in [10]), and k is Boltzmann's constant. When equation 4 is simplified by substituting in the definitions of μ and N_V, letting $s = 2.7$ as in [10], and letting the terms constant with respect to temperature be encompassed by constant C, equation 5 is obtained.

$$\left(\ln\sigma_1 - \ln T^{-1.2}\right) = -\frac{E_A}{k}\frac{1}{T} + \ln C \qquad\qquad (5)$$

Four-point probe conductivity data taken in this work was used to determine the activation energy, E_A, of samples using the relationship of equation 5. To fit the low temperature conductivity, the temperature range used in this study was 22-100 C. An example plot used to determine the activation energy of a sample is shown in Fig. 2.

119

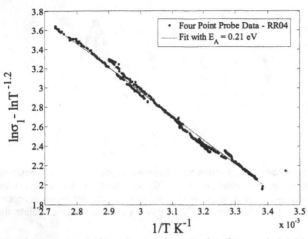

Fig. 2: Temperature dependant conductivity data for sample RR04, fit to determine activation energy.

DISCUSSION

The experimental conditions and the resulting boron incorporation as measured with FTIR are given in Table 1. The diborane concentrations in the gas phase were varied from 1 to 50 ppm in the feedgas. This corresponds to B/C in the plasma discharge being 27-2500 ppm. The boron concentration values in the diamond were determined from the FTIR boron signal at 160 meV for samples RR01-RR04 and at 347 meV for samples 150-154.

Table 1: Experimental deposition conditions and uncompensated boron concentration.

Sample	Pressure (Torr)	CH$_4$ (%)	B$_2$H$_6$ (ppm)	B/C gas phase (ppm)	N$_A$-N$_D$ (cm^{-3})	Thickness (μm)
RR01	135	4	25	1250	1.2×10^{20}	44
RR02	140	4	10	500	3.0×10^{19}	142
RR03	140	4	5	250	1.3×10^{19}	101
RR04	140	4	50	2500	8.7×10^{19}	36
150	160	7.5	1	27	5.6×10^{18}	17
151	160	7.5	2	54	1.1×10^{19}	18
152	130	7.5	1	27	7.8×10^{18}	9
153	140	4	1	50	7.8×10^{18}	18
154	140	4	4	200	2.2×10^{19}	16

The dependence of the boron incorporation versus B/C is shown in Fig. 3. The concentration of N$_A$-N$_D$ determined from the 160 meV absorption and from the 347 meV absorption show increases versus B/C as expected. It is observed that at the same B/C concentration in the gas phase that the incorporation of the boron (N_A-N_D) has some variation between samples. This could be due to a number of factors which include variations of methane concentration, deposition temperature and concentration of defects in the deposited diamond. The boron concentration versus B/C in the source gas from the earlier work of Gheeraert et.al.[5] is also shown on Fig. 3 for comparison. The Gheeraert *et.al.* data is for the absorption at 160 meV and matches well with the data of this study.

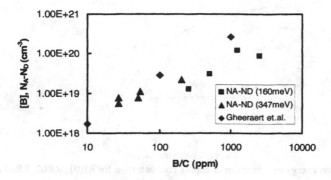

Fig. 3: Uncompensated acceptor concentration versus B/C in gas phase.

The resistivity shows a dependence on the surface treatment of the diamond. Diamond samples are hydrogen plasma treated at the end of the deposition process. After deposition the samples are cleaned in heated nitric and sulfuric acid. It is known that hydrogen can form a surface conduction path [14], can form boron-hydrogen complexes in the diamond [15], and can passivate the electrically active boron near the surface [15,16]. Experiments were also performed that treated the diamond surface with a low pressure oxygen plasma for 3-5 minutes after the initial measurement of the resistivity. In the samples tested, oxygen plasma treated diamond samples showed an increased resistivity as compared to the sample resistivity before oxygen plasma treatment. For example, the resistivity measured for sample 150 went from 8 Ω–cm to 63 Ω–cm after oxygen plasma treatment. The activation energy determined from the temperature dependant conductivity is shown in Fig. 4 for four samples. The activation energy did not change significantly as a result of surface treatment, i.e., comparing pre-oxygen plasma treatment to post-oxygen plasma treatment. The electrical conductivity activation energy varied from 0.27 to 0.21 as N_A-N_D varied from 5×10^{18} to 1×10^{20} cm^{-3}. The evolution of activation energy predicted by the model of Pearson and Bardeen [11] is also plotted for orientation, following the same method as [12].

CONCLUSIONS

Boron doped diamond has been deposited with rates of up to 11.5 µm/hr at higher plasma power densities that occur at pressures of 130-160 Torr. This pressure range is higher than many of the previous studies of boron doped diamond grown by microwave plasma-assisted CVD at pressures below 100 Torr. It should be noted that work by Teraji et.al.[13] was also at higher pressures of 120-135 Torr. The boron was added to the plasma assisted CVD process using diborane gas in amounts of 1-50 ppm in the feedgas of methane and hydrogen. The boron concentration in the diamond varied from 5.6×10^{18} to 1.2×10^{20} cm^{-3} as the diborane increased. At these dopant levels the activation energy of the boron varied from 0.21-0.27 eV, which is less than the 0.37 boron acceptor level at low boron concentrations.

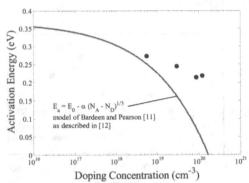

Fig. 4: Activation energy as a function of doping concentration for RR01, RR02, RR04 and seed 150 (•).

REFERENCES
1. R. Ramamurti, M. Becker, T. Schuelke, T. Grotjohn, D. Reinhard and J. Asmussen, *Diam. Rel. Mater.* **17**, 1320-1323 (2008).
2. R. Ramamurti, M. Becker, T. Schuelke, T. Grotjohn, D. Reinhard and J. Asmussen, *Diam. Rel. Mater.* **18**, 704-706 (2009).
3. K.P. Kuo and J. Asmussen, *Diam. Rel. Mater.* **6**, 1097-1105 (1997).
4. A.T. Collins and A.W.S. Williams, *J. Phys. C.: Solid St. Phys.* **4**, 1789-1800 (1971).
5. E. Gheeraert, A Deneuville and J. Mambou, *Diam. Rel. Mater.* **7**, 1509-1512 (1998).
6. K. Thonke, *Semicond. Sci. Technol.* **18**, S20-S26 (2003).
7. V.D. Blank, M.S. Kuznetsov, S.A. Nosukhin, S.A. Terentiev and V.N. Denisov., *Diam. Rel. Mater.* **16**, 800-804 (2007).
8. F.M. Smits, *The Bell System Technical Journal* 711-718 (May 1958).
9. L.B. Valdes, *Proc. IRE*, **42**, 420-427 (1954).
10. J.P. Lagrange, A. Deneuville and E. Gheeraert, *Diam. Rel. Mater.* **7**, 1390-1393 (1998).
11. G.L. Pearson and J. Bardeen, *Phys. Rev,* **75**, 865-883 (1949)
12. T.H. Borst and O. Weis *Diam. Rel. Mater.* **4**, 948-953 (1995).
13. T. Teraji, H. Wada, M. Yamamoto, K. Arima and T. Ito, *Diam. Rel. Mater.*, **15**, 602-606 (2006).
14. K. Hayashi, S. Yamanaka, H. Watanabe, T. Sekiguchi, H. Okushi and K. Kajimura, *J. Appl. Phys.* **81**, 744-753 (1997).
15. C. Baron, M. Wade, A. Deneuville, E. Bustarret, T. Kocinievski, J. Chevalier, C. Uzan-Saguy, R. Kalish and J. Butler, *Diam. Rel. Mater.* **14**, 350-354 (2005).
16. J. E. Butler, M. W. Geis, K. E. Krohn, J. Lawless Jr., S. Deneault, T. M. Lyszczarz, D. Flechtner and R. Wright, *Semicond. Sci. Technol.* **18**, S67-S71 (2003)

Mater. Res. Soc. Symp. Proc. Vol. 1203 © 2010 Materials Research Society   1203-J12-01

Boron Doping in Hot Filament MCD and NCD Diamond Films

Jerry Zimmer[1], Thomas Hantschel[2], Gerry Chandler[1], Wilfried Vandervorst[2,3], Maria Peralta[1]
[1] sp3 Diamond Technologies, 2220 Martin Ave., Santa Clara, CA 95050, U.S.A.
[2] IMEC, Kapeldreef 75, B-3001, Leuven, Belgium
[3]Instituut voor Kern- en Stralingsfysica, K. U. Leuven, Celestijnenlaan 200D, B-3001 Leuven, Belgium

ABSTRACT

Conductive diamond films are essential for electronic applications of diamond but there is still a poor understanding of the effects that growth conditions, grain size and film thickness have on the ultimate conductivity of the film. One of the unique advantages of hot filament diamond is the ability to grow both MCD and NCD films to moderate thicknesses over large areas with little or no change in morphological characteristics such as grain size. In addition the grain size of the film can be altered without the necessity of adding additional gases to the process or unduly increasing the carbon to hydrogen ratio. This gives us an opportunity to investigate electrical conductivity as a function of grain size and thickness within a simple methane, hydrogen, and boron chemical environment over areas which are large enough to support significant production levels of MEMS and other diamond based electronics. In this study the boron source was selected to be trimethyl boron gas to avoid any source of oxygen which could alter the growth conditions and to guarantee that any byproducts of the dopant would be primarily methyl based. The films were grown to various thicknesses up to 5 µm and grain sizes from NCD to full MCD at all thicknesses. This paper explores the effects of both grain size and film thickness on the electrical conductivity of the film as well as the absolute doping levels within the film.

INTRODUCTION

Boron doped diamond grown in microwave plasma deposition systems has been extensively investigated but little has been written about doped diamond grown in hot filament deposition systems. The increasing use of conductive diamond in electrode and micro-electromechanical (MEMS) applications makes it imperative that the entire parameter space of doped diamond be understood to allow volume production of these films with acceptable manufacturing control. May et al. [1] have investigated Raman characteristics of both microcrystalline (MCD) and nanocrystalline (NCD) doped diamond films from hot filament reactors but no one has yet explored the characteristics of films which include the transition region between traditional MCD and NCD regions where average grain size is much smaller than MCD films for a given thickness but the grain size has not yet reached the 100 nm boundary which defines traditional NCD films. The other reason to examine these films is to determine if they are still columnar or are transitioning to a stacked grain structure. Also, minimal work has been done to characterize the full range of dopant levels in all grain size regions. This paper will quantify some of the characteristics in both the MCD region and the transition region to NCD films. Atomic boron levels and electrical conductivity (resistivity) will be correlated to morphology, grain size, and growth conditions for both thin and thick diamond films. The grain sizes are determined using a statistical approach. Furthermore, results will be presented for the possible use of resistivity measurements as a simple but sensitive tool for quantifying grain size and grain size evolution in diamond films.

EXPERIMENTAL PROCEDURE

Diamond films in this work were grown on 100mm diameter silicon wafers using a model 650 hot filament deposition reactor from sp^3 (Santa Clara, California). A simple three gas chemistry of hydrogen (H_2), methane (CH_4), and trimethyl boron (TMB) was used for both MCD and NCD films. No nitrogen, argon, high methane or oxygen bearing gases were used to avoid possible effects of these gases on the conductivity or morphology of the film. Growth condition ranges were nominally 700-900°C substrate temperatures, 2-3% CH_4 levels and 5-30 Torr pressures. Process conditions including temperature, CH_4 level and pressure were adjusted to achieve MCD and near NCD films. TMB flows were adjusted to achieve the full range of conductivity levels from nearly undoped to nearly metallic like.

The sp^3 Model 650 reactor enables the simultaneous deposition on up to nine 100 mm substrates and this allowed samples to be collected at multiple thicknesses by selectively removing samples at different times in the deposition process. Each progressively thicker sample preserves the exact growth conditions of the preceding thinner samples so that grain size evolution can be examined in detail. It also allows a group of samples to be processed through various analytical procedures within a single time window which minimizes any time or calibration dependant variability in the analytical measurements.

The parameter space which was investigated included film thicknesses from 0.5 to 5 μm and atomic boron levels in the films from zero to nearly 2%.

Following deposition, the conductivity and thickness of the films were measured using standard four point probe measurements to obtain sheet resistance and a Filmetrics visible/IR system for the thickness measurement. Samples were then subjected to secondary ion mass spectrometry (SIMS) analysis to measure atomic levels of boron, scanning electron microscopy (SEM) and subsequent image processing to measure the grain size and transmission electron microscopy (TEM) analysis of selected samples to verify grain structure and morphology. Raman measurements were also taken on each sample and will be discussed in reference [2].

DISCUSSION

Boron doping in hot filament diamond can be achieved with a variety of dopant gases including diborane, trimethyl boron, and trimethyl borate. Diborane gas is difficult to handle because of its toxicity and trimethyl borate contains oxygen as part of its structure so the byproducts have a substantial effect on the morphology of NCD films. Trimethyl boron(TMB) is relatively non toxic and pyrophoric in its pure state but is much more manageable when diluted to low levels with hydrogen gas. This makes it the primary choice for creating doped diamond films in a production environment. All films grown for this work used a 2% mix of TMB in hydrogen as a dopant source. Figure 1 shows a typical set of curves for resistivity of MCD and NCD films as a function of TMB:CH_4 gas ratios. Both curves are consistent with dopant levels in other deposition systems where MCD films have significantly higher conductivity than NCD films.

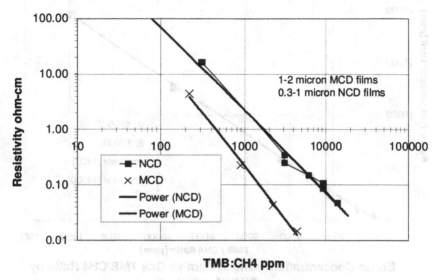

TMB:CH4 ppm

Resistivity vs. TMB:CH4 Ratio

Figure 1

<u>**SIMS Analysis**</u>

The atomic boron concentration in the films was measured by SIMS analysis using known calibration samples measured by elastic recoil detection (ERD). Initially, the boron levels were expected to vary as a function of deposition conditions in addition to the TMB:CH$_4$ ratio. Since the MCD and NCD transitional films were grown under substantially different conditions of temperature and pressure, the data was plotted separately. The results however showed that the atomic boron levels were nearly independent of the range of deposition conditions which were investigated but were determined solely by the TMB:CH$_4$ gas ratio. Figure 2 shows the atomic boron concentration of both types of films as a function of TMB:CH$_4$ gas ratio. It clearly demonstrates that the concentration in the film is a linear function of the concentration in the gas with very little difference in slope between the two film types. It also shows the broad range of boron concentrations that can be achieved with a hot filament reactor.

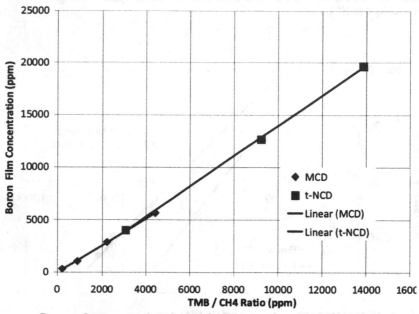

Boron Concentration (ppm) in Film vs Gas TMB/CH4 Ratio by SIMS Analysis

Figure 2

Resistivity

When the four-point-probe sheet resistance data is converted to resistivity the resulting measure of electrically active boron can be compared to the SIMS data, thickness and grain size measurements. The graphs of this data show several interesting tendencies.

Figure 3 shows a plot of resistivity as a function of thickness. The resistivity of the MCD films drops as the thickness increases which is what is expected with columnar structure and increasing lateral grain growth as the film gets thicker. The drop in resistivity starts to saturate at a thickness of about 4 μm however. This suggests that the known effects of grain boundaries on diamond film conductivity are significantly reduced as the grain size increases beyond a few micrometers.

For transitional NCD(t-NCD) films there is a much higher resistivity for any given dopant level as would be expected for a small grain size film. There is a modest drop in resistivity as the film gets thicker which indicates that for the particular growth conditions that were chosen the grain growth is inhibited but not stopped. It may also indicate a columnar structure. The plot is still linear at a thickness of 4-5 μm which suggests that even though grain growth is present the average size is substantially smaller and the rate of increase is slower than in the MCD films.

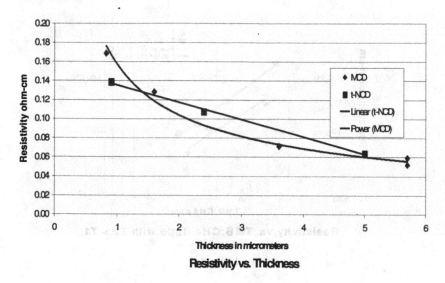

Resistivity vs. Thickness

Figure 3

For true NCD films the grain size should be constant regardless of thickness and the average resistivity should be higher for any given dopant ratio. This can be tested in a hot filament reactor by increasing the growth temperature under conditions favorable to transitional NCD growth. Figure 4 shows the resistivity of films processed at a temperature T2 which is a few 10's of degrees higher than the temperature T1 used for the transitional NCD films. Resistivity is higher for the full range of dopant ratios which is consistent with smaller grain sizes. When these conditions are used to generate films of different thicknesses the resistivity curves for both thicknesses overlap as shown in Figure 5. This indicates that there is very little grain growth for the true NCD films.

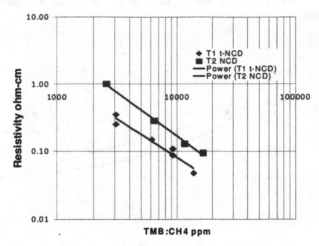

Resistivity vs. TMB:CH4 Ratio with T2 > T1

Figure 4

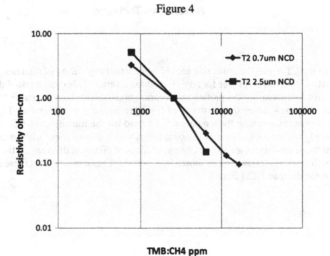

Resistivity vs. TMB:CH4 Ratio - Thin NCD vs Thick NCD

Figure 5

Grain Size

Measurement of average grain size in diamond films presents a number of problems including both the variation in grain size as a function of thickness as well as the distribution of grain sizes at a fixed thickness. In order to accommodate these phenomena, we established a simple grain size measurement procedure which yields both average grain size and grain size distribution.

Surface grain size was selected as an appropriate representation of the overall average film grain size. With that criterion, high resolution SEM images were recorded at a fixed magnification and then subjected to image processing to determine the average grain size of the film. Figure 6 shows typical SEM photographs of 5 micrometer thick t-NCD and MCD films that were used for the measurements. Note that it is important to image a large number of grains to obtain a statistical representative grain size number. It varied in our analysis between 50 and 700 grains for the different images.

Figure 6 – 5 micrometer thick t-NCD (left) and MCD SEM images for grain size calculations

The image processing consisted of using image enhancement to highlight the grain boundaries of the film and then subtracting out the remainder of the photographic image to leave a skeleton rendering of the grain boundaries as shown in Figure 7. In some cases, manual enhancement was used to avoid problems with twin planes causing false grain boundaries to appear in the image.

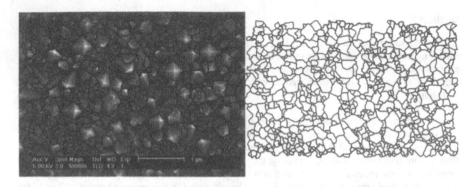

Figure 7 – 0.9 micrometer thick t-NCD SEM with grain boundary enhancement and subsequent background removal

After performing this procedure the image file was further processed using the ImageJ software [3] which is a public domain Java image processing program. It can calculate both area and perimeter information of each shape. The resulting data can then be processed with a standard spreadsheet to obtain the average grain size from both perimeter and area calculations. These can be averaged to obtain an average grain size for each sample as well as a distribution plot of the density of various grain sizes. A typical example of this type of analysis is shown in Figure 8 and Table 1. In addition to the total distribution, the ImageJ data can be manipulated to selected specific ranges of grain size so that unusual effects such as bimodal distributions can be evaluated in detail.

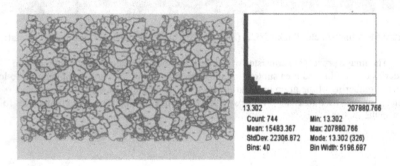

13.302	207880.766
Count: 744	Min: 13.302
Mean: 15483.367	Max: 207880.766
StdDev: 22306.872	Mode: 13.302 (326)
Bins: 40	Bin Width: 5196.687

Figure 8 – Image analysis and grain size(nm^2) histogram from Figure 7 SEM image

Table 1- Grain Size Data

	A=Area [nm²]	P=Perimeter
Average	15483	478
StdDev	22306	368
	$D_1=2\sqrt{(A/\pi)}$	$D_2=P/\pi$
D=Diameter	140	152
Avg	**146 nm**	

Once the grain size can be reproducibly measured then the trends in resistivity which have been ascribed to grain size can be tested for validity. Figure 9 shows the resistivity vs. grain size trends for fixed doping levels and Figure 10 shows grain size vs. thickness for films between 0.5 and 5 μm. Additionally Figure 10 shows the projected grain size for the true NCD films shown in Figure 5 although exact measurements have not yet been made.

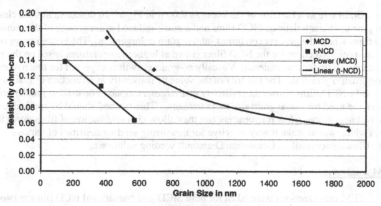

Resistivity vs. Grain Size

Figure 9

The variation in resistivity with grain size for MCD and transitional NCD films follows the same trends as the resistivity vs. thickness plots. MCD films tend to saturate as grain size reaches 1.5 -2 μm and t-NCD films show continuous drops in resistivity as grain size increases but do not saturate because the average grain size is much smaller. In fact, the slope of the t-NCD curve is nearly identical to the slope of the MCD curve at small grain sizes which shows that the two curves would overlap if the absolute doping levels were the same. This in turn implies that the grain size measurement method that was used is consistent across a range of grain sizes and film thicknesses.

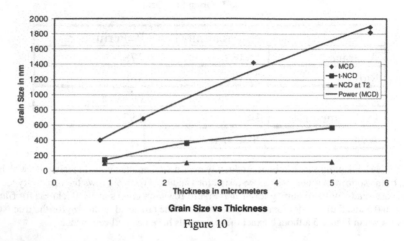

Grain Size vs Thickness
Figure 10

Grain size as a function of thickness as shown in Figure 10 demonstrates the classical grain growth characteristic of MCD films and a much reduced grain growth of the transitional NCD films where the process conditions inhibit grain enlargement. The bottom curve of Figure 10 is the projected curve for the NCD films grown at higher temperatures based on the resistivity vs. thickness data for that material. Virtually no grain growth is expected for this material based on the resistivity results. The linearity of the value of resistivity as a function of grain size for NCD and t-NCD films allows the resistivity measurement alone to be used to predict grain size without the need for actual grain size measurements. This method should be applicable to any film thickness below a few micrometers and may allow characterization of films down to 100 nm or less which would make it very effective for measuring seeding densities of various UDD(Ultrananocrystalline Detonation Diamond) seeding solutions.

TEM Analysis

TEM cross sections were taken for both MCD and transitional NCD films at two different thickness values as shown in Figures 11. The results verify that all of the films show columnar growth regardless of grain size. The growth is columnar in nature to within 10 nm of the interface of the diamond and the substrate as shown in Figure 12. These TEM images verify the predictions of the resistivity measurements that indicated columnar growth based on decreasing resistivity with increasing thickness. They also show that there is minimal if any renucleation of diamond grains during the growth of transitional NCD films. If this result can be extended to NCD films of even smaller grain size such as those grown at higher substrate temperatures then the opportunity exists for making films which have significantly higher Z axis thermal and electrical conductivity. The films would have the mechanical, thermal and electrical properties of large grain material in the vertical (Z) axis while maintaining the smoothness of renucleated NCD films as grown in microwave systems with nitrogen or argon gas chemistry. Further work remains to be done to characterize the NCD films grown at high temperatures in this study and to extend that work to smaller grain sizes. The use of doped films and simple resistivity measurements should make that work possible without the need of extensive TEM or SEM analysis.

Figure 11 – t-NCD Thin and Thick film TEM Cross sections

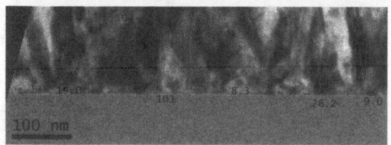

Figure 12 – t-NCD TEM at substrate interface showing columnar growth

CONCLUSIONS

The data presented in this paper shows that boron doped diamond can be deposited with good predictability over a broad range of deposition conditions in a hot filament reactor using a simple three gas chemistry. These films exhibit a linear relationship between atomic boron concentrations and the TMB:CH$_4$ gas ratio with no other first order process dependencies. Electrically active boron concentrations are dependant on the gas ratio but also have a strong dependence on the grain size of the film which is coupled to the thickness of the film for both MCD and transitional NCD films. This thickness dependence is substantially reduced with smaller grain sizes and is not significant in NCD films. TEM analysis shows columnar growth in both MCD and transitional NCD films at thicknesses of 10 nm or more regardless of surface grain size. The columnar growth and lack of renucleation on the transitional NCD films suggests that the vertical electrical and thermal conductivity of these films may be substantially higher than the horizontal values. That aspect of these films may make them attractive for various

MEMS and thin SOI like silicon-on-diamond (SOD) applications where a high vertical thermal conductivity is desirable but lateral conductivity is not important because the diamond is thin.

The apparent lack of sensitivity of the film boron concentration to process conditions makes it possible to consider using a simple resistivity measurement as a tool to measure grain size and grain size evolution in diamond films. This could be used as an in process control system or just to study the effects of various process conditions on grain size for process or product development.

ACKNOWLEDGMENTS

The authors would like to thank Alain Moussa and Lauri Olanterae for grain size analysis; Hugo Bender and Jef Geypen for TEM measurements; Francesca Clemente for RAMAN measurements; and Jozefien Goossens, Joris Delmotte and Bastien Douhard for SIMS analysis.

REFERENCES

1. P.W. May et al. / Diamond & Related Materials 17 (2008) 105–117
2. Hasselt Diamond Workshop 2010 - SBDD XV, February 22-24, 2010 in Hasselt, Belgium
3. ImageJ software available at http://rsbweb.nih.gov/ij/

Surface Modifications

Mater. Res. Soc. Symp. Proc. Vol. 1203 © 2010 Materials Research Society   1203-J17-53

Comparison between chemical and plasmatic treatment of seeding layer for patterned diamond growth

A. Kromka[1], O. Babchenko[1,2], B. Rezek[1], K. Hruska[1], A. Purkrt[1,2], Z. Remes[1]
[1]Institute of Physics AS CR, v.v.i., Cukrovarnicka 10, CZ-16253 Praha 6, Czech Republic
[2]Czech Technical University in Prague, Faculty of Nuclear Sciences and Physical Engineering, Trojanova 13, CZ-12000 Praha 2, Czech Republic

ABSTRACT

We employ UV photolithographic and electron beam lithographic patterning of diamond seeding layer on SiO_2/Si substrates for the selective growth of micrometer and sub-micrometer diamond patterns. Using bottom-up strategy, thin diamond channels (470 nm in width) are directly grown. Differences between wet chemical and plasma treatment on the patterned diamond growth are studied. We find that the density of parasitic diamond crystals (outside predefined patterns) is lowered for gas mixture CF_4/O_2 plasma than for rich O_2 plasma. After CF_4/O_2 plasma treatment, the density of parasitic crystals is 10^6 cm^{-2} which is comparable to the wet chemical treatment. Introducing sandwich-like structure, i.e. photoresist-seeding layer-photoresist, and its treatment (*lift-off* and CF_4/O_2 plasma) further reduces the density of parasitic crystals down to 10^5 cm^{-2}. The advantage of this novel treatment is short processing time, simplicity, and minimal damage of the substrate surface.

INTRODUCTION

An increased interest in nano-crystalline diamond (NCD) films for diverse applications requires also development of techniques suitable for diamond structuring. Such structuring is the key technological step for fabrication of simple as well as sophisticated microelectronic devices [1,2]. Common strategy uses a post-growth plasma treatment in which the fabrication of diamond-based devices consists of at least three technological steps: a) nucleation and/or seeding of non-diamond substrates, b) chemical vapor deposition (CVD) of diamond layer and c) post-growth structuring which is mostly realized by implementing lithographical steps followed by dry etching [3-5]. Limited number of works can be found where very fine and well defined diamond structures (i.e. features far under 1 μm) were realized by reaction ion etching [6]. However, such process is more complex and requires using of plasma-resistive masking material (e.g. SiN_x or Al_2O_3). Finally, the number of technological steps increases and the total processing time is also prolonged.

A selective area deposition (SAD) of NCD structures has been proposed as a promising alternative where the third step, i.e. dry plasma etching of diamond films, is omitted. In SAD, the diamond seeding or nucleation layer is patterned already prior to the diamond growth. This is typically done by photolithographic processing. The SAD was recently achieved also by a direct printing of diamond seeds onto a silicon and quartz substrate with feature sizes of 70 μm [7]. The oldest techniques for the SAD predefined the growth area by a deposition of protection layer (i.e. a masking material used as a sacrificial layer or lift-off layer) or by a material layer which suppressed the diamond formation [8-11]. Fabrication of diamond SAD structures was also achieved by removing the diamond seeds with wet etching (using e.g. buffered oxide etchant – BOE) from areas outside the patterns [12-13]. In general, diverse SAD techniques have been demonstrated by many authors. However, not all of the techniques provide well defined diamond patterns and some are too complex and/or expensive for industrial applications. In addition, the

SAD in sub-micron dimensions (features ≤1 μm) remains still a challenging scientific field. This is due to the fact that in the past, nucleation (seeding) densities were not high enough for growing pinhole-free films and structures, in particular at low layer thickness. The SAD of small structures became feasible once extremely high nucleation densities ($\approx 10^{11}$ cm^{-2}) were achieved by an ultrasonic seeding in nano-sized diamond powder [14,15]. In that case the seeding layer is so densely packed that fully closed layer can be grown as thin as 150 nm [16]. Yet submicron in-plane SAD patterns were not realized so far.

In this paper, we compare two technological strategies in which the growth of micrometer and sub-micrometer diamond patterns is achieved. We show that the strategy No. 1, a wet etching of seeding nano-particles, allows a patterned growth of diamond structures as small as predefined by electron beam lithography. The strategy No. 2 is accomplished by a dry etching of diamond seeds, where a gas composition has a strong influence on removal of seeding nano-particles. We also show that combination of multiple photoresist layers with ultrasonic diamond seeding results in the lowest parasitic seeding density.

EXPERIMENT

Seeding procedure. Si (100) oriented substrates (10x10 mm^2) covered with 1.4 μm thick low temperature oxide SiO$_2$ were ultrasonically cleaned in ethanol (10 min) and dried by a nitrogen gun. Then, they were seeded by a dispersed detonation nanocrystalline diamond powder with a grain size of 5-10 nm (NanoAmando, New Metals and Chemicals Corp. Ltd., Kyobashi) applying an ultrasonic treatment in deionized water for 40 minutes. This resulted in a uniform seeding layer on the substrates. The typical seeding density after such process was up to 10^{11} cm^{-2} [15].

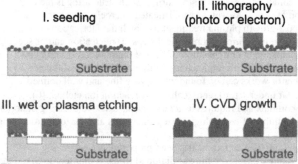

Figure 1. A schematic drawing of 4 basic technological steps involved in used selective area deposition (SAD) strategies.

Treatment of seeded substrates. Schematic illustration of the seeding layer patterning is shown in Fig. 1. In **strategy No. 1**, the seeded substrates were processed by a UV photolithography or electron beam lithography (EBL) process. For photolithography, a photosensitive resin ma-P1215 (phenol-formaldehyde resin) was spin-coated in the thickness of 1.5 μm. This photosensitive polymer was photolithographically processed to predefine growth structures. For the EBL, PMMA resin was spin-coated in the thickness of 60-120 nm. The selectively irradiated photo- and/or EBL resins were developed with an appropriate chemical treatment to form the patterns. The seeding particles localized in the opened regions were removed by immersing the sample in a buffer oxide etchant (BOE). During this step, also SiO_2 layer was partially removed. In **strategy No. 2**, the seeded substrates were covered with a photosensitive OFPR polymer. The OFPR polymer was photolithographically processed to predefine growth structures. Etching of seeds not protected by the polymer was performed by O_2 or CF_4+O_2 r.f. plasma at 100 W under the pressure of 150 mTorr (Phantom LT, Trion Technology). The duration of etching process was 5 min. As an alternative procedure, the Si/SiO_2 substrates are coated by OFPR not only after but also prior to the seeding procedure. Thus sandwich "photoresist-seeding layer-photoresist" structure is formed, patterned by lithography and exposed to plasma.

Diamond CVD growth. In the end of each strategy, the processed samples are loaded into the CVD chamber to realize the CVD growth. The CVD growth was performed in a microwave plasma system (AIXTRON P6) from a methane/hydrogen gas mixture [15]. The process parameters were as follows: 1 % methane in hydrogen, microwave power 1.7÷2.5 kW, total gas pressure 30÷50 mbar and the substrate temperature 580÷800°C. It is important to note that no procedure for removing of OFPR and/or PMMA polymer (the mask) was applied.

Film characterization. The surface morphology of grown structures was investigated by the scanning electron microscope *e_LiNE* system (Raith). For analysis of selectivity the "Atlas" software (Tescan Ltd, Czech Republic) was used. Surface morphology and height of the deposited structures were investigated by atomic force microscopy (AFM) in a tapping mode (AFM Microscope Dimension 3100, Veeco). The set-point ratio (ratio of approached vs. free cantilever oscillation amplitude) was 60%. Silicon AFM cantilevers were used with a typical tip radius of 10 nm and a resonance frequency of 75-230 kHz

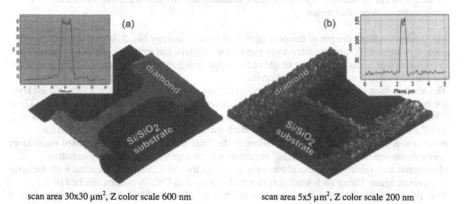

scan area 30x30 µm², Z color scale 600 nm   scan area 5x5 µm², Z color scale 200 nm

Figure 2. AFM images of selective area deposited (SAD) diamond structures using the strategy No. 1 where wet chemical etching of SiO_2 layer was realized by BOE. SAD patterns were fabricated by (a) optical and (b) electron beam lithography.

RESULTS

Fig. 2a shows an AFM scan (30x30 µm²) of selectively deposited diamond channel (5 µm in width and 20 µm in length) by strategy No. 1 using standard photolithographic steps. Total height of the channel structure is 350 nm. However, the diamond channel height is 150 nm while the BOE chemical etching removed 200 nm of SiO_2. Typical grain size, as estimated from SEM images (not shown here), was up to 100 nm. It should be noticed that few isolated objects (clusters of diamond grains) were observed on outside the channel patterns. The density of such "*parasitic*" diamond crystals was approximately 10^6 cm^{-2}. Fig 2b shows the sub-micrometer structures (diamond channels) fabricated by the same strategy No. 1 but using an electron beam lithography. The formed diamond channel structure is 3 µm in length and 470 nm in width. Total height of the channel structure is 150 nm whereas estimated diamond height is approximately 100 nm (the rest represents the etched profile into SiO_2 layer by BOE wet etching). SEM images revealed randomly oriented diamond crystals in sizes up to 80 nm.

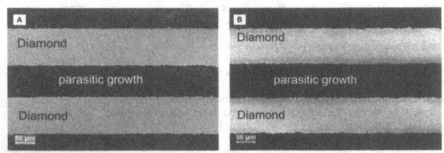

Figure 3. SEM micrograph of diamond patterns (stripes 100 µm width) grown after pre-treatment of (a) the seeding layer/photoresist structure and (b) sandwich-like structure photoresist-seeding-photoresist.

Fig. 3 presents SEM images of samples fabricated by the strategy No. 2. In this case, opened areas (i.e. stripes 100 µm in width) were exposed to reactive ion etching in CF_4/O_2 plasma. The crystal size was up to 750 nm, the height of the stripe was 520÷570 nm. The non-protected areas exhibit randomly distributed diamond crystals.

The areas, which were protected by photoresist during the dry etching, reveal a fully closed layer for both O_2 and CF_4/O_2 plasma types. The main difference between the samples exposed to O_2 or CF_4/O_2 plasma was in the density of parasitic diamond crystals. This density was $9x10^8$ and $8x10^6$ cm^{-2} for O_2 and CF_4/O_2, respectively. Both these values may not be low enough for electronic applications and may cause failure of the final device. As novelty, we used multi-layer "photoresist-seeding layer-photoresist" structure, where a spin coated optically sensitive photoresist was seeded by diamond nano-particles and then this system was coated with the same photoresist again. Using such sandwich structure exposed to CF_4/O_2 plasma resulted in the density of parasitic diamond crystals as low as 10^5 cm^{-2} after the CVD growth, the lowest density reported so far. Table I summarizes parasitic seeding densities for different treatment processes of the patterned seeding layer employed in this study.

Table I. Parasitic density of diamond crystals grown outside predefined patterns after wet chemical and dry plasma treatment of the patterned seeding layer.

process	wet etching	O_2 plasma	$4\%CF_4+O_2$ plasma	2 photoresist layers $4\%CF_4+O_2$ plasma	seeding layer (for comparison)
parasitic density of diamond (nano) crystals	$1x10^6$ cm^{-2}	$9x10^8$ cm^{-2}	$8x10^6$ cm^{-2}	$\approx 10^5$ cm^{-2}	$\geq 10^{11}$ cm^{-2}

DISCUSSIONS

Crucial points in the selective area deposition of diamond structures are the density of parasitic diamond crystals, number of required technological steps and their technological difficulty, and compatibility with large area substrates. In our case, both strategies are compatible with large areas and they are quite simple and inexpensive in principle.

Using the wet chemical treatment (strategy No. 1) resulted in a good selectivity. Parasitic growth was approx. 1 object per 100 μm^2. In our case, the seeding diamond nano-particles are attached to the substrate surface only by the physisorption and their attraction to the substrate can be broken by choosing a proper liquid solution. In addition, part of SiO_2 layer was also etched away in the BOE solution and hence the selective removal of seeding particles was further enhanced. Noticeable is the fact that the diamond microstructures were grown directly on the photoresist mask, as also shown in our previous work [17]. Based on these works and our experimental results we conclude that some amount of photoresist acts as a carbon source which finally accelerates the diamond nucleation and homogenous growth. A detailed study on a comparison of diamond growth on the seeded covered with photoresist is discussed in our previous works [18,19]. The diamond character of the channel structure was confirmed by the Raman mapping. We previously showed that the channel structure exhibited the diamond characteristic line centered at 1332 cm^{-1} while the rest of area was lacking any sign of diamond presence [18]. In addition, these microstructures were successfully employed as p-type conductive channels in the solid-state and solution-gated field effect transistors [20]. Furthermore, using an electron sensitive resist (PMMA) processed by EBL resulted in fabrication of sub-micrometer diamond channels (470 nm in width). This result opens new possibilities in the direct diamond CVD growth for MEMS and sensoric applications. Moreover, the growth of horizontally ordered diamond nano-wire structures seems to be possible by using this SAD strategy avoiding standard problems related to the plasma etching of fully closed diamond layer.

The main advantage of the strategy No.2 is its flexibility and no restriction to use a chemically etchable substrate (like SiO_2). The exposure time during which the masking polymer is not damaged by plasma is one of the main limitations. In our experimental setup, this time period was 5 min for both plasma types. However, applying O_2 plasma etching for 5 min did not remove all the seeding crystals from the exposed areas. Parasitic growth of diamond crystals was observed at relatively high residual densities. On the other side, adding CF_4 to O_2 plasma decreased the residual density down to $8x10^6$ cm^{-2}. Similarly, Ando and his co-workers have observed that adding a small amount of CF_4 to O_2 results in higher concentration and/or reactivity of oxygen ions [21]. Our study indicates that gas mixture of 4%CF_4 in O_2 resulted in more effectively removing of primary (i.e. seeding) diamond nano-particles. The presented residual density ($8*10^6$ cm^{-2}) is comparable with the work of Bongrain et al. [2]. Their residual density was $2*10^7$ cm^{-2} for the exposure time of 5 min. This value dropped down to $5*10^5$ cm^{-2} for the prolonged exposure time of 20 minutes. In our case, we observed that a prolonging of the exposure time over 5 minutes resulted in a damaging of masking polymer and porous diamond structures were formed. It should be noticed that Bongrain et al. used a sacrificial metal mask which is able to withstand longer exposure times. We avoided the use of metal mask by implementing the multi-layer structure photoresist-seeding layer-photoresist treated in CF_4/O_2 plasma. The density of parasitic diamond crystals decreased down to $\approx 10^5$ cm^{-2}. This value nearly corresponds to inherent density of surface defects, i.e. energetically favorable sites for stimulating diamond growth. It should be noticed that parasitic nano-particles were homogenously distributed over the whole area. A slight increase in their concentration was observed at the edges of patterns.

CONCLUSIONS

In this paper we employed two technological strategies for the selective area deposition of diamond structures. Sub-micrometer diamond channel in width of 470 nm were fabricated by an employing the EBL process and wet etching of the seeding layer on SiO_2. This strategy seems to be suitable for fabrication of horizontally aligned diamond nano-wires and/or nano-gaps. The second strategy was based on dry etching of seeding diamond nano-crystals using reactive ion etching through the polymer mask. Adding of 4% CF_4 to oxygen increased the efficiency in etching of diamond seeds resulting in a low value of the parasitic density (8×10^6 cm^{-2}). Further decreasing of the residual density by increasing the etching time was not possible while the masking polymer did not withstand the etching process. Implementing novel multi-layer "photoresist-seeding layer-photoresist" structure, processed by UV optical lithography (*lift-off* process) followed by CF_4/O_2 plasma etching, decreased the residual density down to physical limit of 10^5 cm^{-2}. The highlight of this *multilayer* structure is its efficiency, simplicity and minimal damage of the substrate surface. It is believed that combining this technique with EBL will progress the highly selective diamond growth into a sub-micrometer size region.

ACKNOWLEDGEMENT

This work was supported by the Academy of Sciences of the Czech Republic grants IAAX00100902, KAN400100701 and KAN400480701, by the Institutional Research Plan No. AV0Z10200521 and project No. LC-510, and by the J.E.Purkyne Fellowship.

REFERENCES

1 Y. Fu, H. Du, J. Miao, *Jour. Mat. Proc. Technol.* **132**, 73 (2003).
2 A. Bongrain, E. Scorsone, L. Rousseau, G. Lissorgues, C. Gesset, S. Saada and P. Bergonzo, *J. Micromech. Microeng.*, **19**, 074015 (2009).
3 G.F. Ding, H.P. Mao, Y.L. Cai, Y.H. Zhang, X. Yao, X.L. Zhao: *Diam. Relat. Mater.* **14**, 1543 (2005)
4 M.P. Hiscocks, C.J. Kaalund, F. Ladouceur, S.T. Huntington, B.C. Gibson, S. Trpkovski, D. Simpson, E. Ampem-Lassen, S. Prawer, J.E. Butler: *Diam. Relat. Mater.*, **17**, 1831 (2008).
5 Y. S. Zou, Y. Yang, W. J. Zhang, Y. M. Chong, B. He, I. Bello, S. T. Lee, *App. Phys. Lett.* **92**, 053105 (2008).
6 H. Gamo, K. Shimada, M. Nishitani-Gamo, T. Ando: Jpn. J. Appl. Phys. **46**, 6267 (2007).
7 N.A. Fox, M.J. Youh, J.W. Steeds, W.N. Wang, *J. Appl. Phys.* **87**, 8187 (2000).
8 Y.-H. Chen, C.-T. Hu, I-N. Lin: *Jap. Jour. Appl. Phys* **36**, 6900 (1997).
9 Y. Sakamoto, M. Takaya, H. Sugimura, O. Takai, N. Nakagiri: *Thin Solid Films*, **334**, 161 (1998).
10 A. Massod, M. Aslam, M.A. Tamor, T.J. Potter, *J. Electrochem. Soc.* **138**, L67 (1991).
11 M.W. Geis, J.C. Twichell, T.M. Lyszczarz, *J. Vac. Sci. Technol.* **B14** 2060 (1996).
12 K. Hirabayashi, Y. Taniguchi, O. Takamatsu, T. Ikeda, K. Ikoma and N. Iwasaki-Kurihara: *Appl. Phys. Lett.* **53**, 1815 (1988).
13 H. Liu, C. Wang, C. Gao, Y. Han, J. Luo, G. Zou and C. Wen: *J. Phys.: Condens. Matter*, **14**, 10973 (2002).
14 O. A. Williams, O. Douhéret, M. Daenen, K. Haenen, E. Osawa, M. Takahashi, *Chem. Phys. Lett.* **445**, 255 (2007).
15 A. Kromka, B. Rezek, Z. Remes, M. Michalka, M. Ledinsky, J. Zemek, J. Potmesil, M. Vanecek, *Chemical Vapor Deposition* **14**, 181 (2008).

16 A. Kriele, O.A. Williams, M. Wolfer, D. Brink, W. Muller-Sbert, C. E. Nebel, *Appl. Phys. Let.* **95**, 031905 (2009).

17 A. Kromka, O. Babchenko, B. Rezek, M. Ledinsky, K. Hruska, J. Potmesil, M. Vanecek, *Thin Solid Films* **518**, 343 (2009).

18 A. Kromka, O. Babchenko, H. Kozak, K. Hruska, B. Rezek, M. Ledinsky, J. Potmesil, M. Michalka, M. Vanecek: *Diam. Relat. Mater.* **18**, 734 (2009).

19 A. Kromka, O. Babchenko, H. Kozak, B. Rezek and M. Vanecek, *phys. stat. sol. b* **246** 2654 (2009).

20 H. Kozak et al, accepted for publication in *Sensor Letters* (2009).

21 Y. Ando, Y. Nishibayashi, K. Kobashia, T. Hirao, K. Oura: *Diam. Relat. Mater.* **11**, 824 (2002).

Mater. Res. Soc. Symp. Proc. Vol. 1203 © 2010 Materials Research Society   1203-J17-44

Interaction of hydrogen and oxygen with nanocrystalline diamond surfaces

Thomas Haensel[1], Syed Imad-Uddin Ahmed[1], Jens Uhlig[1], Roland J. Koch[1], José A. Garrido[2], Martin Stutzmann[2], and Juergen A. Schaefer[1,3]
[1]Institut für Physik and Institut für Mikro- und Nanotechnologien, Technische Universität Ilmenau, 98693 Ilmenau, Germany
[2]Walter Schottky Institute, Technical University Munich, 85748 Garching, Germany
[3]Department of Physics, Montana State University, Bozeman, Montana 59717, USA

ABSTRACT

Nanocrystalline diamond films (NCD) are strong candidates for applications in a wide variety of fields. An important concern in all these applications is to understand the properties of variously prepared NCD surfaces. This contribution is focussed on the surface science study of hydrogen and oxygen containing NCD films using X-ray photoelectron spectroscopy (XPS) as well as high resolution electron energy loss spectroscopy (HREELS). Previous studies have demonstrated that hydrogen, oxygen, and gases from the ambient environment as well as water can result in drastic surface changes affecting conductivity, wettability, tribological properties, etc. In this contribution we analyzed differently prepared NCD surfaces as a function of parameters such as the annealing temperature under ultrahigh vacuum conditions (UHV). We are able to identify the thermal stability of a number of species at the interface, which are related to different characteristics of C-H, C-OH, C=O, and C=C bonds. Furthermore, a formation of graphitic-like species appears at higher annealing temperatures. An atomic hydrogen treatment was also applied to the NCD surface to obtain further information about the surface composition.

INTRODUCTION

Diamond and related materials like diamond-like carbon are of great interest as transparent protective coatings, wear-resistant coatings, electronic devices, sensor systems, and bioelectronic systems. It was found that the graphitic-like species in amorphous carbon are responsible for the electrical behavior [1,2] while the diamond phase is important for the hardness. Furthermore, hydrogen bonded on the diamond surface can cause negative electron affinity (NEA), which might be of interest for electron emitting materials. In contrast, additional oxygen bonded to the surface can cause a positive electron affinity (PEA) [3]. Therefore, drastic changes of the electronic behavior can occur depending on the functional groups present on the surface. However, due to their electrical properties, nanocrystalline diamond is also used as a biosensor material [4]. Single crystalline diamond films are hard to produce and still very expensive. In most cases nanocrystalline diamond (NCD) is the first choice as these films can be manufactured on various substrates with a low roughness and good electrical properties [1,2] that can be controlled with appropriate doping [5]. Since in most applications diamond films are used under ambient conditions, fundamental research on the reactivity with hydrogen and oxygen of NCD surfaces is needed. Also, the functionalization with biomolecules [1] is an interesting aspect for the use of NCD. In this case it is important to know which functional group is produced as a result of surface treatments such as annealing or oxidation.

Using X-ray photoelectron spectroscopy (XPS) it is possible to measure the carbon and oxygen content in the films. Furthermore, as the binding energies of the graphitic-like and the diamond-like phase are shifted, it is possible to differentiate these two phases with a fitting procedure of the C1s core levels in the NCD. High resolution electron energy loss spectroscopy (HREELS) is a well known technique that is extremely surface sensitive to vibrational modes as well as plasmons. Therefore, HREELS can be used to measure the fingerprints of NCDs, which are also correlated to the grain size [6]. Comparing HREELS and XPS it is possible to obtain an understanding of the functional groups present on the surface.

Due to the fact that hydrogen and oxygen are the main chemical reaction species of NCD films, an oxygen-plasma treatment and also additional a subsequent hydrogen-treatment was applied to get specific surface conditions. Afterwards, the surface conditions were analyzed with HREELS and XPS.

EXPERIMENTAL DETAILS

The NCD-films were grown on Si substrates using plasma CVD in an Ar-atmosphere. Nitrogen was added into the gas phase and, therefore, should be conductive. The oxidation process was performed in an O-plasma in an Asher for 5min with 200W. Two kinds of surface treatments were applied in this study. The first was treated in an O-plasma only (designated from now on as O-NCD) while the other was treated in the same O-plasma and subsequently treated for 30 min with radical H atmosphere created by a hot-wire setup (designated from now on as H-NCD). The as-grown samples were cleaned in solvents. The O-NCD and H-NCD samples were later analyzed without further preparation. After preparing the samples they were transferred into an UHV vacuum setup with a base pressure of 2×10^{-10} Torr. In this vacuum system it was possible to anneal the sample in the pressure range below 10^{-9} Torr up to 1400 °C from the rear sample side with an e-beam heating system.

The samples were analyzed using a HREELS spectrometer (SPECS Delta 0.5) with an adjusted resolution of 2.5 meV or 5 meV. XPS measurements were performed with a non-monochromatic Mg Kα X-ray source (V = 12 kV, I = 20 mA, hν = 1253.6 eV, X-ray line width 0.7 eV).

RESULTS AND DISCUSSION

HREELS

Using HREELS measurements, it is possible to analyze dipole vibrations of adsorbed and bonded groups as well as plasmons. Spectra of clean NCD films are well known and can be used as fingerprints [6]. The loss spectra of the as-loaded samples can be seen in figure 1. The intensity ratio of the 1st multiple (1M) νC-C and/or δC-H mode (at about 2420 cm^{-1}) to the νC-H mode (at about 2880 cm^{-1}) changes depending on the grain size [6]. The estimated grain size of the used NCD films is approximately 5-10 nm. Comparing the spectra with other measurements the fingerprints and the v C-H to 1M intensity ratio of these films are in good agreement with literature [6].

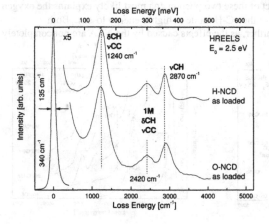

Figure 1. HREELS spectra of the as-loaded O-NCD and H-NCD samples. The FWHM of the quasi-elastically reflected electrons and the vC-H mode (at about 2900 cm^{-1}) are higher for the oxygen-rich NCD compared to H-NCD.

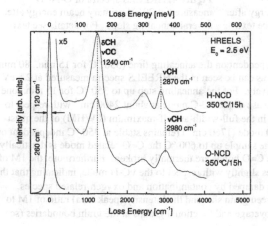

Figure 2. HREELS spectra of O-NCD and H-NCD after annealing up to 350 °C for about 15 hours. The energy loss of the vC-H mode (about 2900 cm^{-1}) and the FWHM of the quasi-elastically scattered electrons are still higher for O-NCD compared to H-NCD.

Furthermore, the vC-H mode is shifted for the O-NCD sample with respect to the H-NCD one. Due to the fact that as-loaded samples are often contaminated with a thin water film or a thin carbon contamination layer [7], annealing steps up to 350 °C were applied and the changes are shown in figure 2. However, this shift of the oxygen treated NCD was measured again. This suggests that this shift might occur due to graphitic-like phases at the grain boundaries, as the vC-H modes are at higher energy losses for the graphitic-like species [6, 7]. However, an oxygen-induced effect has also to be taken into account. The FWHM of the quasi-elastically

147

reflected electrons is correlated with the charge density, which is lower for the O-NCD sample (table 1). It is known that smaller FWHM values are achieved for conducting compared to the non-conducting materials. Also, a higher graphitic-like carbon content is responsible for higher charge densities. The combined effect of these two phenomena most likely explains the oxygen-induced effect that results in shifting of the vC-H mode to higher energy losses. Both interpretations need to be analyzed further, as the effects caused by the NEA are not completely understood.

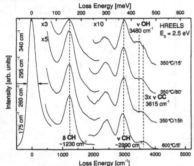 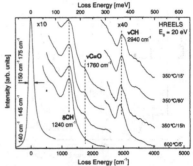

Figure 3. HREELS spectra of O-NCD measured at 2.5 eV primary beam energy after several annealing steps. For details see text.

Figure 4. HREELS spectra of O-NCD measured at 20 eV primary beam energy after several annealing steps. For details see text.

Desorption of contaminants is also dependent on the annealing time (350 °C for 15 min., 80 min. and 15 hours, and 600 °C for 5 min.) as can be seen in the HREELS spectra measured at 2.5 eV and 20 eV in figures 3 and 4, respectively. For the annealing step up to 350 °C for 15 min., one can see a desorption of vO-H, an increase of the vC-C mode at about 2400 cm^{-1} with respect to the vC-H mode as well as a decrease in the full-width-at-half-maximum (FWHM) of the quasi-elastically reflected peak. The vC=O mode (1760 cm^{-1}) remains stable at 350 °C independent of annealing time (Fig. 4). Annealing the sample up to 600 °C the C=O related mode is drastically decreased (Fig. 4), indicating that the C=O bonds are thermally broken. Furthermore, the 1M of the vC-C and/or δC-H mode increases slightly with respect to the vC-H mode, indicating that the 1M with respect to the vC-H mode is damped by contamination and oxygen-related species. This implies that the correlation between grain size and the intensity (peak area) ratio of 1M to vC-H is also dependent on oxygen-coverage and functional groups at the grain boundaries (see table 1).

Upon annealing the sample up to 900 °C for 5 minutes the vC-H mode is drastically decreased compared to annealing steps up to 600 °C, indicating that hydrogen is desorbed. This phenomenon is well known [7]. Furthermore, the O-NCD sample has a higher FWHM relative to the H-NCD of the quasi-elastically scattered electron peak, indicating changes in the charge densities [8,9] and the vC-H mode is lower. Thus, a stable regime for hydrogen bonding does not exist at 900 °C for both, H-NCD and O-NCD. As the oxygen-content is zero for both samples, one should expect nearly the same HREELS spectra for both samples. The fact that they are still different can be explained by the fact that the C-O bonded states are removed during the annealing procedure and a difference in the graphitic-like phases and even C=C reconstructions might be the reasons for the changes in the structural and electrical properties of both annealed NCD samples.

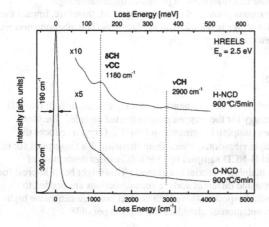

Figure 5. HREELS spectra of O-NCD and H-NCD after annealing up to 900 °C for 5 min. FWHM of elastically scattered electrons is higher and νC-H is negligible compared to H-NCD.

XPS

XPS data were taken in conjunction with HREELS to be certain that no contamination occurs during annealing as well as for the identification of the carbon and oxygen content by O1s and C1s peak area determinations (Tab. 1). For this analysis, the cross sections from oxygen and carbon were taken from Yeh and Lindau [10] and the analyzer transmission was assumed to be proportional to $1 / E_{kin}$, as is usual for a hemispherical analyzer. Comparing the results of the HREELS measurements and the XPS data there is a clear correlation between the oxygen content of the NCD sample with the FWHM of the quasi-elastically scattered electrons.

Table 1: Carbon and oxygen content in the NCD samples estimated with XPS for a homogeneous mixture compared to the FWHM of the quasi-elastically reflected electrons and the intensity ratios of the νC-H to 1M peaks.

Sample	Preparation	O [at%] : C [at%]	FWHM [cm^{-1}]	$I_{νC-H} / I_{1M}$
O-NCD	as-loaded	15 : 85.0	335	1.8
	350 °C 15 min	6.5 : 93.5	340	2.45
	350 °C 80 min	5 : 95.0	295	2.25
	350 °C 15 h	3 : 97.0	160	2.00
	600 °C 5min	0.3 : 99.7	175	1.35
	900 °C 5 min	0 : 100	298	-
H-NCD	as-loaded	1 : 99.0	135	0.95
	350 °C 15 h	0.5 : 99.5	120	0.70
	650 °C 5 min	0 : 100	135	0.70
	900 °C 5min	0 : 100	190	-

A decrease of the oxygen content of the O-NCD sample due to the annealing procedure causes a decrease of the peak area ratio of the νC-H to 1M peaks, indicating that the oxygen content as well as the grain-size are important for that ratio.

In tribology it is well established that high contact pressures often cause high temperatures in the contact zone. The results presented in this study could, therefore, impact the chemical conversion of carbon based coatings in tribological applications. This phenomenon will be examined more closely in further studies.

SUMMARY AND CONCLUSIONS

The results of the HREELS and XPS measurements show a significant shift of the C-H stretching vibration towards higher energy for the oxygen plasma treated sample. The C=O bonds are still stable at 350 °C, and are completely removed at 600 °C. Oxygen reduces the intensity ratio of the νC-H to 1st multiple vibrations, which from literature is a fingerprint for the grain-size [6]. Annealing O-NCD and H-NCD samples up to 900 °C causes desorption of oxygen and hydrogen, and then the remaining material is different. This might be of interest for surface functionalization, where defects and different surface reconstructions are needed to attach molecules, and also for tribological aspects, where high contact pressure can cause high contact temperatures and therefore an initiation of chemical reactions is possible.

ACKNOWLEDGMENTS

This work was supported in part by the German Research Foundation (DFG) within the framework of the Collaborative Research Center 622 "Nanopositioning and Nanomeasuring Machines". Technical support by G. Hartung is gratefully acknowledged.

REFERENCES
[1]   J. Robertson, Phys. Stat. Sol. A **205**, 2233 (2008)
[2]   D. M. Gruen, Annu. Rev. Mater. Sci. **29**, 211 (1999)
[3]   S. Torrengo, L. Minati, M. Filippi, A. Miotello, M. Ferrari, A. Chiasera, E. Vittone, A. Pasquarelli, M. Dipalo, E. Kohn, G. Speranza, Diamond Relat. Mater. **18**, 804 (2009)
[4]   M. Stutzmann, J. A. Garrido, M. Eickhoff, M. S. Brandt, Phys. Stat. Sol. A **203**, 3424 (2006)
[5]   P. Achatz, J. A. Garrido, M. Stutzmann, O. A. Williams, D. M. Gruen, A. Kromka, and D. Steinmüller, Appl. Phys. Lett. **88**, 101908 (2006)
[6]   S. Michaelson, O. Ternyak, R. Akhvlediani, and A. Hoffman, Chem. Vap. Deposition **14**, 196 (2008)
[7]   T. Haensel, J. Uhlig, R. J. Koch, S. I.-U. Ahmed, J. A. Garrido, D. Steinmüller-Nethl, M. Stutzmann, J. A. Schaefer, Phys. Stat. Sol. A **206**, 2022 (2009)
[8]   H. Ibach and D. L. Mills, Electron Energy Loss Spectroscopy and Surface Vibrations, Academic Press, New York, 1982
[9]   T. Balster, V.M. Polyakov, H. Ibach, and J.A. Schaefer, Surface Science **416**, 177 (1998)
[10]   J. J. Yeh, I. Lindau, At. Data Nucl. Data Tables **32**, 1 (1985)

Mater. Res. Soc. Symp. Proc. Vol. 1203 © 2010 Materials Research Society

CH$_2$ group migration between the H-terminated 2×1 reconstructed {100} and {111} surfaces of diamond

J.C. Richley, J.N. Harvey and M.N.R. Ashfold
School of Chemistry, University of Bristol, Bristol, BS8 1TS, United Kingdom

ABSTRACT

Various possible routes for the migration of a CH$_2$ group between the H-terminated 2×1 reconstructed {100} surface and the H-terminated {111} surface of diamond have been explored using a hybrid quantum mechanical/molecular mechanical method. The calculated energies suggest that movement of such surface bound species across step edges should be a facile process under typical diamond growth conditions, and that such migrations are significant contributors to the observed morphologies of diamond grown by chemical vapor deposition methods.

INTRODUCTION

Chemical vapor deposition (CVD) techniques for growing diamond films from an activated hydrocarbon/hydrogen gas mixture are now well established [1,2]. Activation of the gas mixture – with, for example, a hot filament, or in a microwave plasma – results in H$_2$ dissociation and production of H atoms. These H atoms react with the hydrocarbon source gas via a series of addition and/or abstraction reactions, creating a wide variety of carbon containing species – both radicals and stable molecules. Gas phase H atoms can also abstract surface terminating H atoms (*i.e.* from surface C–H bonds), thereby creating temporary radical sites on the growing diamond surface. The most probable fate of these radical sites will be re-termination by another gas phase H atom, but occasionally they will bond with an incident carbon-containing radical (*e.g.* a CH$_3$ radical). Subsequent H atom abstraction (creating a pendant CH$_2$ group), surface rearrangement and addition steps result in eventual incorporation of the incident carbon atom into the diamond lattice.

Migration of surface bound carbon radical species on the 2×1 {100} surface [3-7] and on the {111} surface [8] have both been studied previously. In both instances, migration requires that the pendant CH$_2$ group reacts with an adjacent surface radical site, forming a (strained) ring; subsequent re-opening of this ring results in, either, reversion to the starting configuration, or movement of the CH$_2$ group along the diamond surface by one carbon atom. Such migrations have been advanced as making an important contribution to the smooth surface morphology that is often observed in diamond samples grown by CVD [5]. Migration from one diamond surface to another (*i.e.* across a step edge) has received less attention, but may also be important in determining the morphology of as-grown CVD diamond. The aim of the present study is to investigate the energetics of migration between the H-terminated 2×1 reconstructed {100} and the {111} surfaces of diamond.

THEORY

The present calculations employ hybrid quantum mechanical/molecular mechanical (QM/MM) methods. The calculations were performed using the QoMMMa program [9,10]; calculations for the QM region were performed using Jaguar 5 [11], while the MM region was modeled using TINKER [12]. The geometry of the QM region was optimized using the B3LYP density functional with the 6-31G(d) basis set. The MM region was described using the MM2 protocol. As in the previous studies [7], the starting species for migration involve a pendant –CH_2 group, and a neighboring radical site, and hence have two unpaired electrons. The final species also have two unpaired electrons, whereas the intermediate strained-ring systems have closed-shell electronic structures. All reported energies are provided relative to the triplet electronic state of the starting species, described using an unrestricted DFT ansatz. The corresponding open-shell singlet, again modeled with an unrestricted ansatz, lies within ~10 kJ mol^{-1} in energy of the triplet. The transition states for ring-opening and closing were studied using unrestricted DFT, with the ring-closed species described using restricted closed-shell DFT.

Two principal types of step edge can be formed by juxtaposing a {111} surface between two 2×1 {100} surfaces. These are henceforth termed convex and concave, as illustrated in fig. 1.

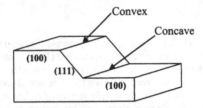

Figure 1. Illustration of the two step edges created between the {100} and {111} surfaces.

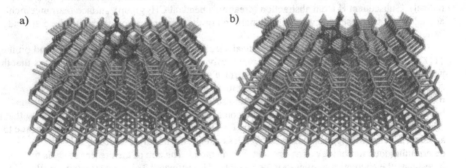

Figure 2. Models used to simulate the convex step edge between the 2×1 {100} and {111} diamond surface: a) dimer rows perpendicular b) dimer rows parallel to the step edge. The region treated with MM methods is shown in grey, while the blue region in this illustration is treated using QM.

Two models have been used to describe the convex step edge between the 2×1 H-terminated {100} and the H-terminated {111} surfaces. Both are based on suitably modified versions of the 5×9×4 slab (defined in terms of the numbers of C–C dimer bonds) used in our earlier studies of CH₃ addition to the 2×1 {100} diamond surface [7]. These differ in the relative orientations of the dimer rows on the upper 2×1 {100} surface, which can be either parallel or perpendicular to the step edge as shown in fig. 2.

There are two possible types of intersection between the {111} and {100} surfaces at the concave step-edge. Given that for one of these topologies, there are two distinct routes from the {111} surface leading down to the {100} surface, we needed to consider three possible pathways for migration across the concave step-edge. The corresponding QM regions are shown in fig. 3.

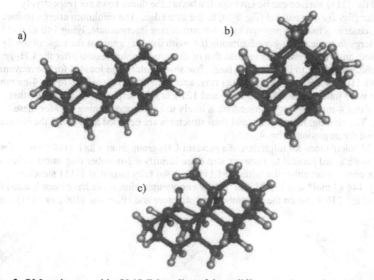

Figure 3. QM regions used in QM/MM studies of three different pathways by which a CH_2 group might migrate from a {111} face to a 2×1 {100} terrace at a concave step-edge.

Note that these QM regions are considerably larger than most considered in our earlier migration study [7] and, at this stage, we have only determined approximate transition states for the various CH_2 migration processes by calculating the energy of the system along a specified reaction coordinate. The system is held close to the desired value of this reaction coordinate using a harmonic constraint. QM and MM energies calculated for each value of the reaction coordinate are combined and the transition state identified as the maximum energy along this curve. All QM/MM energies reported are in kJ mol⁻¹ relative to the triplet diradical starting species.

RESULTS AND DISCUSSION

Migration of a CH_2 group on the diamond surface starts with the abstraction of a surface hydrogen species from a carbon atom adjacent to the radical group of interest. The CH_2 group can then move between the radical sites via a ring closing and subsequent ring opening mechanism. The present calculations start with a pendant CH_2 group adjacent to a surface radical site (*i.e.* the earlier H abstraction steps have been skipped) and attention is focused on the ring closing / ring opening sequence that enables migration of the CH_2 group.

Convex step edge

As fig. 2 showed, two different scenarios for CH_2 migration between the upper 2×1 {100} surface and the {111} surface can be envisaged, wherein the dimer rows are respectively perpendicular (fig. 2(a)) or parallel (fig. 2(b)) to the step edge. The minimum energy pathway for the former scenario is found to proceed via a 4-member ring intermediate, lying 140 kJ mol^{-1} higher in energy than the starting configuration (*i.e.* with the CH_2 group at the edge of the 2×1 {100} surface) and 109 kJ mol^{-1} higher than that of the structure that results after the CH_2 group has successfully migrated to the {111} surface. Transition states were located for the movement from the 2×1 {100} surface to the 4-member ring, and from the {111} surface to the 4-member ring. These were found to lie at 166 kJ mol^{-1} and 121 kJ mol^{-1}, respectively, indicating that formation of this 4-member ring intermediate is likely to be the rate limiting step for these migrations. The starting, intermediate and final structures are depicted along with the relevant energy profile for migration in fig. 4(a).

QM/MM calculations for migration of a pendant CH_2 group from a 2×1 {100} surface with the dimer rows aligned parallel to a convex step edge identify a 3-member ring intermediate that lies lower in energy than either the initial 2×1 {100} or the fully migrated {111} structures – by, respectively, 146 kJ mol^{-1} and 175 kJ mol^{-1}. Low energy transition states have been located for the movement of the CH_2 between the 3-member ring structure and either the {100} or {111} surface.

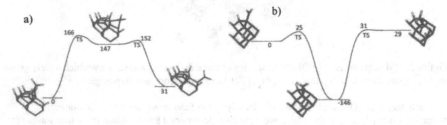

Figure 4. Energy profiles for the migration of a CH_2 group from the 2×1 {100} surface to the {111} surface. a) dimer rows perpendicular to the step edge b) dimer rows parallel to the step edge. Energies shown are in kJ mol^{-1} and are defined relative to the initial structure with the migrating CH_2 group located on the 2×1 {100} surface.

Concave step edge

Migration of a CH$_2$ group from the {111} surface to the 2×1 {100} surface via a concave step edge can proceed via three different routes, as implied by the different QM regions displayed in fig. 3. The dimer rows on the receiving 2×1 {100} surface may be either perpendicular (fig. 3(a)) or parallel to the step edge. The latter case allows further variety, since the migrating CH$_2$ group can approach from either between the dimer chains (fig. 3(b)) or from the middle of a dimer (fig. 3(c)). These different pathways are henceforth referred to as P1, P2 and P3, respectively. Optimized geometries and energies for the CH$_2$ group located on the {111} surface, on the 2×1 {100} surface and in a bridging ring geometry have been calculated for each of the three pathways. The calculated energy differences, defined relative to the initial structure with the pendant CH$_2$ group located on the {111} surface, are shown in Table 1.

Table 1. Relative energies of the fully migrated structure with the CH$_2$ group on the 2×1 {100} surface ($\Delta E_{\{100\}}$), and of the 3-member bridging ring intermediate (ΔE_{ring}), respectively, defined with respect to the initial structure with the pendant CH$_2$ group located on the {111} surface. The calculations are at the B3LYP/6-31G(d):MM2 level of theory, and the quoted energies are in kJ mol^{-1}.

Pathway	$\Delta E_{\{100\}}$	ΔE_{ring}
P1	-99	-363
P2	-33	-119
P3	-60	-171

Pathways P2 and P3 proceed via similar energy intermediate 5-member ring structures, which enable facile transfer of the CH$_2$ group from the {111} surface to the 2×1 {100} surface, and vice versa. The energy minimum associated with the 6-member ring bridging structure in pathway P1 is much deeper, however, suggesting that once a migrating CH$_2$ group reaches this bridging position between the 2×1 {100} and {111} surfaces it is very likely to remain fixed at this location – *i.e.* to incorporate into the diamond lattice at the step edge, in accord with the earlier conclusions from Frenklach *et al* [6].

CONCLUSIONS

QM/MM computational methods have been used to explore the migration of a surface bound CH$_2$ species between the 2×1 {100} and {111} surfaces of diamond at two different step edges. The resultant energies suggest that migration of a CH$_2$ group between these surfaces will be a reasonably facile process at typical substrate temperatures (T_{sub} ~1000-1400 K), and that such migrations are thus likely to be a significant factor in determining the observed morphologies of as-grown CVD diamond.

ACKNOWLEDGMENTS

We are grateful to EPSRC, for the reward of a research grant and studentship, and to Element Six Ltd. for financial support.

REFERENCES

[1] D. G. Goodwin and J.E. Butler in *Handbook of Industrial Diamonds and Diamond films*, ed. M.A. Prelas, G. Popovici, L.G. Bigelow. (Marcel Dekker: New York, 1998) pp.527-81.

[2] J.E. Butler, Y.A. Mankelevich, A. Cheesman, J. Ma and M.N.R. Ashfold, *J. Phys.: Condens. Matter* **21**, 364201 (2009).

[3] M. Frenklach and S. Skokov, *J. Phys. Chem. B.,* **101**, 3025 (1997).

[4] S. Skokov, B. Weiner and M. Frenklach, *J. Phys. Chem,.* **88,** 7073 (1994).

[5] S. Skokov, B. Weiner, M. Frenklach, Th. Frauenheim and M. Sternberg, *Phys. Rev. B.,* **52**, 5426, (1995).

[6] M. Frenklach, S. Skokov and B. Weiner, *Nature*, **372**, 535 (1994).

[7] A. Cheesman, J.N. Harvey and M.N.R. Ashfold, *J. Phys. Chem. A.*, **112**, 11436 (2008).

[8] K. Larsson and J.O. Carlsson, *Phys. Rev. B.*, **59**, 8315 (1999).

[9] J.N. Harvey, *Faraday Discuss.*, **127**, 165 (2004).

[10] A.C. Tsipis, A.G. Orpen and J.N. Harvey, *Dalton Trans.*, 2849 (2005).

[11] Jaguar, Schrödinger Inc., Portland, OR, (2000).

[12] J.W. Ponder, TINKER: Software Tools for Molecular Design, v4.0: St. Louis, MO (2003).

Characterization

Mater. Res. Soc. Symp. Proc. Vol. 1203 © 2010 Materials Research Society   1203-J17-13

Optical monitoring of nanocrystalline diamond with reduced non-diamond contamination

Z. Remes, A. Kromka, T. Izak, A. Purkrt and M. Vanecek

Institute of Physics of the ASCR, v.v.i., Cukrovarnicka 10, Praha 6, Czech Republic

ABSTRACT

Previously, the nanocrystalline grain boundaries were often contaminated by the "non-diamond phase" with the photo-ionization threshold at 0.8 eV. Here, we present the optical spectra of the NCD films grown on transparent substrates by the microwave plasma enhanced chemical vapor deposition (CVD) at a relatively low temperature below 600°C. The transmittance and reflectance spectra are useful to evaluate the film thickness, the surface roughness and the index of refraction. The direct measurement of the optical absorptance by the laser calorimetry and photothermal deflection spectroscopy (PDS) provides high sensitive methods to measure the weak optical absorption of thin films with rough surface. The optical measurements indicate the high optical transparency of our standard, nominally undoped 0.2-0.3 μm thick NCD film with low non-diamond content. However, the optical scattering is rather high in UV and needs to be reduced.

INTRODUCTION

In the last years a progress in the microwave plasma enhanced chemical vapor deposition (MW PECVD) allowed to prepare nanocrystalline diamond (NCD) layers with a optical high quality [1, 2]. Nanocrystalline diamond (NCD) films with typical grain size below 100 nm are relatively smooth showing most of the diamond excellent properties such us optical transparency in wide range, excellent mechanical properties, bio-compatibility and chemical inertness. The nanodiamond surface can be controllably modified for applications [3-5].

At the Institute of Physics of the ASCR, v.v.i. in Prague we grow nominally undoped NCD films with grain size below 100 nm on a wide range of substrates using the microwave plasma enhanced chemical vapour deposition (MW CVD) [6]. Our NCD films show high dark resistivity and measurable UV photosensitivity after surface oxidation [7]. Previous studies showed usefulness of photothermal deflection and spectrally resolved photocurrent spectroscopy to estimate sub-gap defect density. In particular, the onset of absorption at about 0.8 eV in undoped films attributed to transitions from π to π* states introduced into the band gap by the high amount of sp^2 bonded carbon at the grain boundaries has been reported [8]. In our previous papers we have applied optical methods to analyze the optical absorption in a broad spectral range as well as the surface roughness. We have shown that the non-diamond content can be reduced by several orders of magnitude by depositing NCD on the carefully selected UV-grade fused silica substrates under the optimized growth conditions followed by the post-deposition chemical etching and cleaning [9]. In this paper we discuss in detail the optical properties of our nominally undoped nanodiamond films grown on fused silica substrates. In particular we compare the transmittance & reflectance spectroscopy (T&R), photothermal deflection spectroscopy (PDS) and laser calorimetry.

EXPERIMENTAL DETAILS

The continuous, optically smooth, UV transparent and nominally undoped NCD films were grown in the Aixtron P6 MW PECVD reactor from a hydrogen rich CH_4 /H_2 gas phase on high purity fused silica glass substrates seeded in the ultrasonic bath with dispersed nanodiamond powder [10-12]. Before the deposition process, the substrates were ultrasonically cleaned by isopropanol and deionized water and seeded in ultrasonic bath using a nanodiamond powder (5 nm). The depositions were carried out with the following parameters: hydrogen gas flow 300 sccm, methane gas flow 3 sccm, vacuum pressure 30 mbar, deposition power 2500 W, substrate temperature 800°C. The post-growth acid treatment was performed in order to reduce the amount of non-diamond phase the films by boiling NCD in acid solution (H_2SO_4+KNO_3) at 200°C for 30 min and then rinsed in deionized water [13].

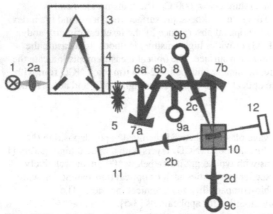

Figure 1 The schematic view of our T&R and PDS spectrometer: 1) Xe lamp, 2a-d) lenses, 3) monochromator, 4) optical filters, 5) chopper, 6a-b) flat mirrors, 7a-b) off-axis parabolic mirrors, 8) beam-splitter, 9a-c) detectors, 10) sample immersed in transparent liquid, 11) He-Ne laser, 12) position detector

Optical properties of NCD films deposited on transparent substrate have been investigated by optical transmittance, reflectance and PDS in our laboratory-made spectrometer, see Fig.1. The samples are immersed in the transparent liquid and all optical spectra are measured simultaneously without moving the sample. The spectrometer works in the dual-beam mode to enhance reproducibility. The monochromatic light is focused in our standard measurements onto small spot 2 x 0.5 mm. The optical spectra are measured with dynamic range by 4 orders of magnitude down to 0.01% in the spectral range 200-2200 nm. The spectral range is limited by transparency of the immersion liquid, by transparency of cuvette and by the spectral sensitivity of the detectors. In our standard measurements we use as a detector an integration sphere equipped with photodiodes. The compound UV-enhanced Si and InGaAs detector allows very precise T&R measurements in the spectral range 200-1700 nm.

Laser calorimetry (LC) was measured with 0.01°C precision at room temperature (22°C) and pressure 1 Pa inside an Optistat DN-V Oxford Instruments cryostat using a type E thermocouple and solid state continuous lasers operating at 532 nm, 830 nm and 1064 nm. The sample was insulated from the copper holder by glass slide with the thermocouple pressed between the sample and the glass slide minimizing the thermal contact between the sample and the holder. The thermocouple voltage was recorded every 1 s and recalculated to the temperature difference between the sample and the copper holder. The intensity of the laser light was measured by the spectrally calibrated Si photodiode. The incident light intensity was corrected to reflectance of the cryostat window.

DISCUSSION

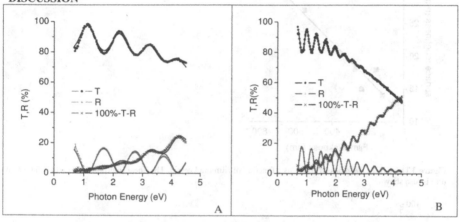

Figure 2 The specular transmittance T and reflectance R spectra of thin NCD films deposited on fused silica glass and immersed in CHCl$_3$. The solid curves represent the theoretical fits A) $n=2.33 + 0.014/\lambda^2$, λ in μm, $d=236$ nm, $\sigma=15$ nm and B) $n=2.34 + 0.013/\lambda^2$, λ in μm, $d=790$ nm, $\sigma=33$ nm, calculated under assumption $A=0$.

The specular transmittance and reflectance spectra of our typical NCD films are shown in Fig. 2a and 2b. The optical spectra show interference fringes which are very sensitive to the film thickness d and surface roughness. Since T and R spectra were measured at exactly the same spot, without moving the sample, it was possible to evaluate the spectral function 100%-T-R. The reflectance between substrate and transparent liquid can be neglected in the case of the optical matching between the substrate and the transparent liquid surrounding the sample. Therefore, the glass/CHCl3 interface has negligible reflectance. In order to evaluate the surface scattering, we applied the theoretical expressions for the optical reflectance R and transmittance T for a free-standing thin film incorporating the effects of the multi-reflections, interference and surface scattering at one surface where the reduction of the amplitudes of the reflected and transmitted light is expressed in terms of optical phase shifts which depend on the rms surface roughness σ [14]. We achieved a good agreement between the measured and calculated T&R spectra just fitting the film thickness d, the surface roughness σ and two parameters in the Cauchy formulas for the index of refraction $n = n_0 + n_1/\lambda^2$ (λ is wavelength in μm). Fig. 2a and 2b show that T and R spectra can be fitted by four parameters (n_1, n_2, d and σ) with high accuracy supposing negligible optical absorptance A in the region below the diamond optical absorption edge. The spectral function 100%-T-R then represents the surface scattering. Although the immersion of the rough TCO into the transparent liquid reduces the surface scattering, the surface roughness significantly deteriorates optical spectra in UV region. Nevertheless, NCD film is still partly transparent even in UV region.

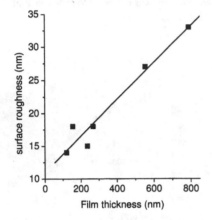

Figure 3 The surface roughness σ (rms) as a function of film thickness d. The linear interpolation $\sigma = 11 + 0.03d$ (in nm) is also shown.

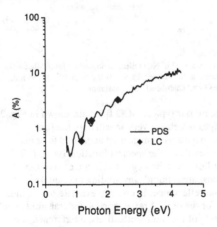

Figure 4 The optical absorptance A of 0.8 μm thick NCD film measured by photothermal deflection spectroscopy (PDS) and put in the absolute scale by laser calorimetry (LC) measured at selected wavelengths.

The average size of diamond nanocrystals increases in our as grown films with film thickness increasing the surface roughness, see Fig 3. The optical scattering on nanorough surface dominates the spectral function $100\%\text{-}T\text{-}R$ and it prevents from evaluating the small optical absorption in a visible and near IR range directly from the transmittance and reflectance spectra. Although the immersion of the rough TCO into the transparent liquid reduces the surface scattering, the surface roughness significantly deteriorates optical spectra in UV region for thick NCD layers, see Fig. 2.

It is difficult to evaluate the optical absorptance A from $100\%\text{-}T\text{-}R$ spectra if $A < 1\%$ even in the case of smooth layer. On the other hand, photothermal methods provide direct and sensitive tools for measuring the optical absorption. The direct measurement of the optical absorptance spectra has been done by photothermal deflection spectroscopy (PDS) where the deflection of the probe beam is measured by the position detector. The probe beam is parallel and close to the sample surface. Its deflection is caused by thermal waves induced in the transparent liquid surrounding the sample by periodical heating of the sample by amplitude modulated monochromatic light (pump beam). The PDS spectra were normalized on the incident monochromatic light intensity by spectrally calibrated detector and set on the absolute optical absorptance scale using the optical absorptance measured by the laser calorimetry, see Fig 4. Laser calorimetry allows at selected wavelengths an absolute estimate of the optical absorptance. Assuming a direct conversion of absorbed energy to heat, the optical absorptance A can be calculated from the detected temperature rise $\Delta T(t)$ according to formulas described in Ref. [15] where the incident laser power, mass of the substrate and specific heat capacity of the substrate needs to be know and two assumptions must be fulfilled. First, the light is absorbed only in NCD film. Second, since the NCD film is in a good thermal contact with the substrate, the heat generated in the NCD film is quickly dissipated into the substrate keeping both NCD and substrate the same temperature.

N-incorporation in CVD diamond is very easy even for very low N concentrations in the gas phase, as it can be measured by optical emission spectroscopy (OES) of the plasma during microwave CVD of diamond thin films with nitrogen addition [16]. Typically the NCD samples contain N in various forms. Even if the absorption of the substitution N is low, additional bands are present as NV-, NV0 and the band at 4.5 eV so all N-related absorption must be taken into account[17]. Nevertheless, N-related optical absorptance is negligible in IR region whereas Fig. 4 shows the measurable optical absorptance in all spectra range. Thus, the optical absorptance spectrum in Fig. 4 is dominated by the residual non-diamond inclusions due to the optical transitions between sp$_2$-bonded carbon atoms which introduce in diamond the π and π^* states located symmetrically around the Fermi energy in the band gap [18].

CONCLUSIONS

The $T\&R$ spectra are useful to evaluate the film thickness d, the surface roughness σ and the index of refraction (Cauchy parameters n_0 and n_1). The surface scattering increases with increasing the surface roughness and decreasing wavelength causing the discrepancy between $100\%\text{-}T\text{-}R$ and optical absorptance spectra in VIS and UV region. The direct measurement of the optical absorptance by the laser calorimetry and photothermal deflection spectroscopy (PDS) provides highly sensitive methods to measure the weak optical absorption of thin films with rough surface. The optical spectroscopy indicates the high optical transparency of our standard,

0.2-0.3 µm thick NCD film with low non-diamond content. However, the optical scattering is rather high and it needs to be reduced.

ACKNOWLEDGEMENT
This work was supported by the Academy of Sciences of the Czech Republic grant IAAX00100902, by the Institutional Research Plan No. AV0Z10200521, project No. LC-510, and by the Fellowship J.E.Purkyne.

REFERENCES

[1]   D. M. Gruen, *Annual Review of Materials Science* **29,** 211(1999).
[2]   O. A. Williams, M. Nesladek, M. Daenen, S. Michaelson, A. Hoffman, E. Osawa, K. Haenen and R. B. Jackman, *Diamond and Related Materials* **17,** 1080 (2008).
[3]   S. Wenmackers, S. D. Pop, K. Roodenko, V. Vermeeren, O. A. Williams, M. Daenen, O. Douheret, J. D'Haen, A. Hardy, M. K. Van Bael, K. Hinrichs, C. Cobet, M. Vandeven, M. Ameloot, K. Haenen, L. Michiels, N. Esser and P. Wagner, *Langmuir* **24,** 7269 (2008).
[4]   J. Cermak, B. Rezek, A. Kromka, M. Ledinsky and J. Kocka, *Diamond and Related Materials* **18,** 1098 (2009)
[5]   L. Michalikova, B. Rezek, A. Kromka and M. Kalbacova, *Vacuum* **84,** 61 (2009).
[6]   A. Kromka, B. Rezek, Z. Remes, M. Michalka, M. Ledinsky, J. Zemek, J. Potmesil and M. Vanecek, *Chemical Vapor Deposition* **14,** 181 (2008).
[7]   Z. Remes, A. Kromka, M. Vanecek, S. Ghodbane and D. Steinmuller-Nethl, *Diamond and Related Materials* **18,** 726 (2009).
[8]   P. Achatz, J. A. Garrido, M. Stutzmann, O. A. Williams, D. M. Gruen, A. Kromka and D. Steinmuller, *Applied Physics Letters* **88,** 101908 (2006).
[9]   Z. Remes, A. Kromka and M. Vanecek, *Physica Status Solidi (A) Applications and Materials* **206,** 2004 (2009).
[10]   S. Potocky, A. Kromka, J. Potmesil, Z. Remes, V. Vorlicek, M. Vanecek and M. Michalka, *Diamond and Related Materials* **16,** 744 (2007).
[11]   O. A. Williams, O. Douheret, M. Daenen, K. Haenen, E. Osawa and M. Takahashi, *Chemical Physics Letters* **445,** 255 (2007).
[12]   A. Kromka, S. Potocky, J. Cermak, B. Rezek, J. Potmesil, J. Zemek and M. Vanecek, *Diamond and Related Materials* **17,** 1252 (2008).
[13]   H. Kozak, A. Kromka, E. Ukraintsev, J. Houdkova, M. Ledinsky, M. Vanecek and B. Rezek, *Diamond and Related Materials* **18,** 722 (2009).
[14]   Z. Yin, H. S. Tan and F. W. Smith, *Diamond and Related Materials* **5,** 1490 (1996).
[15]   U. Willamowski, D. Ristau and E. Welsch, *Applied Optics* **37,** 8362 (1998).
[16]   T. Vandevelde, M. Nesladek, K. Meykens, C. Quaeyhaegens, L. M. Stals, I. Gouzman and A. Hoffman, *Diamond and Related Materials* **7,** 152 (1998).
[17]   K. Iakoubovskii and G. J. Adriaenssens, *Journal of Physics Condensed Matter* **12,** L77 (2000).
[18]   M. Nesladek, K. Meykens, L. M. Stals, M. Vanecek and J. Rosa, *Physical Review B - Condensed Matter and Materials Physics* **54,** 5552 (1996).

Mater. Res. Soc. Symp. Proc. Vol. 1203 © 2010 Materials Research Society   1203-J17-06

Modification of the electrical and optical properties of single crystal diamond with focused MeV ion beams

E. Vittone[1], O. Budnyk[1], A. Lo Giudice[1], P.Olivero[1], F. Picollo[1], Hao Wang[1], F. Bosia[2], S. Calusi[3], L. Giuntini[3], M. Massi[3], S. Lagomarsino[4], S. Sciortino[4], G. Amato[5], F. Belotti[5], S. Borini[5], M. Jaksic[6], Ž. Pastuović,[6], N. Skukan[6], M. Vannoni[7]

[1] Experimental Physics Department / Centre of Excellence "Nanostructured Interfaces and Surfaces", University of Torino, and INFN sez. Torino Via P. Giuria 1, 10125 Torino, Italy
[2] Department of Theoretical Physics, University of Torino, and INFN sez. Torino, Italy
[3] Department of Physics, University of Firenze and INFN sez. Firenze, Italy
[4] Department of Energetics, University of Firenze and INFN sez. Firenze, Italy
[5] Quantum Research Laboratory, Istituto Nazionale di Ricerca Metrologica, Torino, Italy
[6] Laboratory for Ion Beam Interactions, Ruđer Bošković Institute, Zagreb, Croatia
[7] CNR Istituto Nazionale di Ottica Applicata (INOA), Firenze, Italy

ABSTRACT

In this paper an overview is given on recent results obtained in the framework of an Italian/Croatian collaboration aimed to explore the potential of techniques based on focused MeV ion beams to locally modify the structural, electrical and optical features of diamond.

Experiments were carried out using light (H, He, C) ion beams with energies of the order of MeV, focused to micrometer-size spot and raster scanned onto the surface of monocrystalline (IIa or Ib) diamond samples. Different energies, ion species and fluences were used, in conjunction with variable thickness masks and post annealing processes, to define three-dimensional structures in diamond, whose electrical/optical/structural properties have been suitably characterized. Finite element numerical methods have been employed in the modeling of the material modification and in device design.

INTRODUCTION

The extreme properties of diamond make this material appealing for many applications, ranging from ionizing radiation detectors to bio-sensors, from optical and photonic devices to micro-fluidic and electromechanical systems [1].

The full exploitation of the vast potential of this material requires a fine modification of the structural, optical and electrical properties of synthetic diamond single crystals, which represent the ideal substrates in terms of material quality and reproducibility.

The use of MeV ions represents one effective approach to achieve this goal [2], since the damage induced by ion beam irradiation enables the modification of the physical properties of diamond with high spatial resolution, both in depth (by tuning the end-of-range damage profile with sub-micrometer accuracy using different ion species and energies) and in the lateral directions (by focusing the ion beams down to micrometer-sized spots). A high control on the induced damage density is achieved by careful monitoring of the implantation fluence.

At high damage levels, the conversion to a graphite-like phase allows the definition of electrically conductive paths to be employed as buried electrodes. At lower damage densities, diamond retains its basic structural properties, while significantly changing its optical properties (namely, refractive index and absorption).

In this paper, we summarize recent results obtained by our collaboration to characterize the structural/electrical/optical properties of monocrystalline diamond samples, previously damaged with MeV focused ions beams, as function of their fluences and energies.

EXPERIMENT

The experiments were carried out using commercially available synthetic single crystal parallelepiped-shape diamond samples, produced by Sumitomo (type Ib) and Element Six (type IIa), with thicknesses of 1.5 and 0.5 mm, respectively. The two opposite large faces are optically polished and oriented along the (100) crystallographic axis.

Implantations were carried out at several scanning ion beam facilities, namely: the Legnaro National Laboratories (LNL) [3], the Ruđer Bošković Institute (RBI) [4,5] and the "Laboratory of nuclear techniques for cultural heritage" (LABEC) [6,7]. Various ion species and energies (namely 6 MeV C, 2 and 3 MeV H, 1.8 MeV He) were employed at fluences ranging from 10^{13} to 10^{17} ions cm^{-2}. The beams were focused to spot dimensions of the order of few micrometers and raster scanned on the polished side of the samples.

The morphology of the implanted regions were analyzed both by white light profilometry (WLP) [6] and Atomic Force Microscopy (AFM) [4]; the relevant refractive index modifications were determined with a laser interferometric microscope [7].

RESULTS

Fig. 1 shows the average vacancy profile generated in diamond by the above mentioned ions, as evaluated by SRIM [8] simulations. Even though the simulations ignores some effects (i.e. channeling, self-annealing, etc.), it gives a sufficient insight of the structural damage caused by ion implantation; in particular, the profile indicates that the damage extends along the whole ion range, and presents a maximum at the end of range. By scanning the focused ion beam along pre-defined paths, amorphous and graphite-like regions can be produced with micrometric resolution below the surface down to a depth depending on the energy and type of ions, as shown in the optical micrograph in Fig. 2 (left).

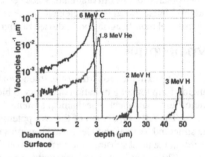

Figure 1: SRIM evaluations of the average vacancy profiles in diamond by 6 MeV C, 1.8 MeV He and 2 MeV/3MeV H ions.

Structural modifications

The amorphisation of diamond caused by ion bombardment induces a structural modification of the material as a consequence of the remarkable density variation of the damaged regions ($\rho_{aC} \approx 1.6$ g cm^{-3}) with respect to the crystalline surroundings ($\rho_D \approx 3.5$ g cm^{-3}). This

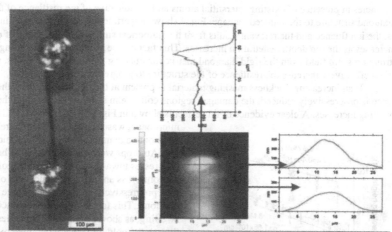

Figure 2: (left) Optical images in reflection of a diamond sample after irradiation with 6 MeV C focused ion beam. The opaque strip-like structures are the graphitic-like damaged regions. The dark rectangles are the Cr/Au pads where gold drops are deposited (bright structures). (right) AFM topography map at the endpoint where the damaged region is connected with the sample surface; the arrows indicate the directions along which the profiles are collected. The increasing swelling thickness at the channel endpoint is clearly visible.

entails a volume expansion, which is constrained in the vertical direction and gives rise to surface swelling [7]. Fig. 3 shows the topography of a 350×350 μm^2 region after the irradiation of a 125×125 μm^2 region with 2 MeV H ions at a fluence of $7.6\cdot10^{16}$ cm^{-2} as measured by WLP. The analysis of the surface swelling was carried out by numerical simulations based on the Finite Element Method [7]. The density ρ of the damaged regions as function of the ion fluence F and the linear density of vacancies $\lambda(z)$ (see Fig. 1) is modeled with the following formula:

$$\rho(F,z) = \rho_d - \left(\rho_D - \rho_{aC}\right)\cdot\left(1 - e^{-\frac{F\lambda(z)}{\alpha}}\right)$$

where α is a free parameter which is varied to get the best agreement with the experimental data. It expresses the probability of creating additional

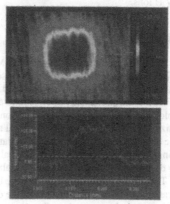

Figure 3: (top) Topography of a sample irradiated with a fluence of 7.63 10^{16} cm^{-2} 2 MeV protons. Fieldmap in false colours of the swelling on the surface, corresponding to an under-surface implantation; (bottom) profile of the sample surface along the line shown in the top image.

vacancies in presence of existing interstitial atoms and is indicative of the resilience of the diamond structure to the induced damage. Fig. 4 shows experimental data of the swelling heights vs. the ion fluence and the relevant results from the numerical simulation. The value of α increases as the ion depth penetration increases. This fact can be interpreted considering the strong pressure field from the rigid diamond matrix surrounding the implanted regions, which could effectively increase the resilience of the structure to progressive amorphization.

If an increasing thickness masking material is present at the diamond surface, the ion range is progressively reduced, the damaged regions collapse in shallower layers and the surface swelling increases. A clear evidence of this effect is shown in Fig. 2 (right); a 6 MeV C ion microbeam was scanned along strip areas 16 µm wide onto a diamond surface where micrometer-size Au drops were deposited. As the beam scan progresses towards the center of the gold drop, the ions cross an increasing thickness of gold, thus progressively reducing their range in diamond..This fact increases the surface swelling, as shown by the AFM profiles. Even though the gold drops were obtained by means of a standard gold wire ball bonder and the shape and dimensions of these first variable-thickness masks (diameter of around 50 µm) were not optimized, these preliminary results provide promising evidence that the use of variable thickness masks is suitable to fabricate three dimensional structures embedded in a diamond matrix.

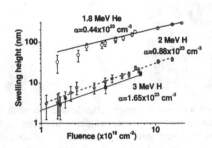

Figure 4: Experimental (markers) and numerical (lines) swelling values h vs. fluence F for 1.8 MeV He, 2MeV H and 3MeV H ions.

Electrical modifications

The bombardment of diamond with MeV ions provokes the transition from an electrically highly insulating material to a conductive pseudo-graphitic phase through the progressive generation of a network of defective sp^3 and sp^2 sites [9].

As the damage density goes beyond a critical threshold (referred as "graphitization threshold"), a continuous network of sp^2 bonded defects is formed, which leads to the permanent graphitization of the structure upon thermal annealing and to the subsequent appearance of metallic conductivity. At a lower level of damage, or in the absence of any annealing process following the ion implantation, the electrical conduction is similar to what observed in a-C and usually described by models based on variable range hopping mechanism [10,3].

These two conduction regimes have been studied in ion damaged regions of diamond, using variable thickness-masks to make the damaged regions more easily accessible to connections with the external reading circuit [3,4].

Figure 5 shows the dependence of the conductance G in the ohmic regime of a strip-like region in diamond irradiated with 1.8 MeV He ions at a fluence of $5.2 \cdot 10^{17}$ cm^{-2} as function of the temperature. The "$T^{-1/4}$" dependence, characteristic of the variable range hopping in the localized states near the Fermi level, is clearly evident in the 270-690 K temperature range. From the slope of the linear fit, the density of states at the Fermi level is estimated as $N(E_F)=5.5 \cdot 10^{17}$

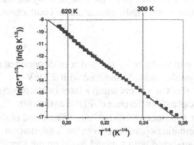
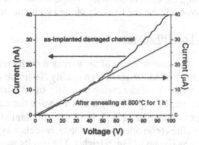

Figure 5: (left) Plot of $\ln[G(T) \cdot T^{1/2}]$ vs $T^{-1/4}$ in the ohmic regime of regions irradiated with 1.8 MeV He ions at a fluence of $5.2 \cdot 10^{17}$ cm^{-2}. The line is the linear fit in the temperature range 270-690 K. (right) IV characteristics of as implanted and annealed strip-like regions damaged with 6 MeV C ions at a fluence of $1.1 \cdot 10^{16}$ cm^{-2}.

cm^{-3} eV^{-1}. A comparison of such result with others available in literature highlights that the relationship between $N(E_F)$ and the damage is influenced not only by the vacancy density, and hence by the ion fluence, but also by the ion range, i.e. by the ion type and energy. This fact can be ascribed to the resilience of the diamond lattice to structural damage at different depths and is in agreement with the conclusions drawn from the analysis of the structural modification occurring in ion damaged diamond, as described in the previous section.

Thermal annealing following ion implantation can cause both the recovery of the lightly damaged regions and the permanent graphitization of the damaged layers, whose vacancy density goes beyond the graphitization threshold. In fig. 5 (right), the current-voltage characteristic of a strip-like damaged region, as obtained by the irradiation with 6 MeV C ions at a fluence of $1.1 \cdot 10^{16}$ cm^{-2}, before and after the annealing at 800 °C for 1 hour in vacuum. Before

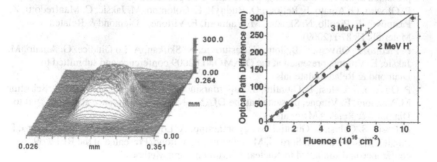

Figure 6: (left) example of the optical path difference map measured with the laser interferometric microscope; the area of interest was implanted with 2 MeV protons at a fluence of 7.63×10^{16} cm^{-2}. (right) Experimental (dotted) and fitted (line) plots of the optical path difference as a function of implantation fluence of 2 MeV and 3 MeV protons.

annealing, the current behavior is superlinear whereas, after annealing, the conductivity increases of a factor 10^3 and the linearity of the IV characteristics clearly indicates an ohmic conduction.

Optical modifications

The measurement of the variation of the real part of the refractive index of diamond at a wavelength of 632.8 nm in high quality IIa single crystal diamond irradiated with 2 MeV protons with fluences ranging from 10^{13} to 10^{17} cm^{-2} was carried out using a laser interferometric microscope [6]. Fig. 6 (left) shows a map of the optical path difference (OPD) measured in correspondence of a 125×125 μm^2 area implanted at a fluence of $7.6 \cdot 10^{16}$ cm^{-2}. The trend of the OPD profiles (corrected for swelling), reveals a systematic dependence from the implantation fluence, as shown in Fig. 6 (right). The results were analyzed with a model based on the specific damage profile (Fig. 1) and on the assumption of a linear dependence of the refractive index from the vacancy density; the continuous curves in fig. 6 are the best fit of the experimental data,

From the consistency between data relevant to 2 and 3 MeV protons, the model is assumed to be applicable for any damage profile generated by ions with different species and energies, provided that the damage density does not exceed $\sim 3 \cdot 10^{21}$ vac. cm^{-3}; the maximum variation of the refractive index amounts to $\sim 4\%$ of the absolute value at $\lambda = 632$ nm (n=2.41).

ACKNOWLEDGMENTS

This work is supported by the "Accademia Nazionale dei Lincei – Compagnia di San Paolo" and by "FARE" experiments of INFN, which are gratefully acknowledged.

REFERENCES

1. R. W. Jackman, Semicond. Sci. Technol. 18, S1-S140 (2003)
2. R. Kalish et al., Nucl. Instr. Methods in Phys. Res. B 106, 492 (1995)
3. P. Olivero, G. Amato, F. Bellotti, S. Borini, A. Lo Giudice, F. Picollo, E. Vittone, presented at the EMRS2009 conference and to be published in the Eur. Phys. Jour. B.
4. P. Olivero, G. Amato, F. Bellotti, O. Budnyk, E. Colombo, M. Jakšic, C. Manfredotti, Ž. Pastuovic, F. Picollo, N. Skukan, M. Vannoni, E. Vittone, , Diamond & Related Materials 18 870 (2009).
5. F. Picollo, P. Olivero, F. Bellotti, Ž. Pastuović, N. Skukan, A. Lo Giudice, G. Amato, M. Jakšić, E. Vittone, presented at the DIAMOND 2009 conference and submitted to Diamond & Related Materials.
6. P. Olivero, S. Calusi, L. Giuntini, S. Lagomarsino, A. Lo Giudice, M. Massi, S. Sciortino, M. Vannoni, E. Vittone, presented at the DIAMOND 2009 conference and submitted to Diamond & Related Materials.
7. F. Bosia, S. Calusi, L. Giuntini, S. Lagomarsino, A. Lo Giudice, M. Massi, P. Olivero, F. Picollo, S. Sciortino, A. Sordini, M. Vannoni, E. Vittone, Presented at the REI2009 conference and submitted to Nuclear Instruments and Methods B.
8. J.F. Ziegler, J.P. Biersack, M.D. Ziegler, *"SRIM, the stopping and range of ions in matter"*, SRIM Co., MA, USA, www.SRIM.org.
9. J. F. Prins, Phys. Rev. B 31, 2472 (1985)
10. D. Saada, J. Adler, R. Kalish, Int. J. Mod. Phys., C 9 (1998) 61.

Device Applications

(1) Bio-Applications

Mater. Res. Soc. Symp. Proc. Vol. 1203 © 2010 Materials Research Society   1203-J17-31

Capacitive field-effect (bio-)chemical sensors based on nanocrystalline diamond films

M. Bäcker[1,2], A. Poghossian[1,2], M. H. Abouzar[1,2], S. Wenmackers[3], S. D. Janssens[3], K. Haenen[3,4], P. Wagner[3,4], and M. J. Schöning[1,2]

[1]Institute of Nano- and Biotechnologies, Aachen University of Applied Sciences, Campus Jülich, Germany
[2]Institute of Bio- and Nanosystems, Research Centre Jülich, Germany
[3]Institute for Materials Research, Hasselt University, Diepenbeek, Belgium
[4]Division IMOMEC, IMEC vzw., Diepenbeek, Belgium

ABSTRACT

Capacitive field-effect electrolyte-diamond-insulator-semiconductor (EDIS) structures with O-terminated nanocrystalline diamond (NCD) as sensitive gate material have been realized and investigated for the detection of pH, penicillin concentration, and layer-by-layer adsorption of polyelectrolytes. The surface oxidizing procedure of NCD thin films as well as the seeding and NCD growth process on a Si-SiO$_2$ substrate have been improved to provide high pH-sensitive, non-porous thin films without damage of the underlying SiO$_2$ layer and with a high coverage of O-terminated sites. The NCD surface topography, roughness, and coverage of the surface groups have been characterized by SEM, AFM and XPS methods. The EDIS sensors with O-terminated NCD film treated in oxidizing boiling mixture for 45 min show a pH sensitivity of about 50 mV/pH. The pH-sensitive properties of the NCD have been used to develop an EDIS-based penicillin biosensor with high sensitivity (65-70 mV/decade in the concentration range of 0.25-2.5 mM penicillin G) and low detection limit (5 µM). The results of label-free electrical detection of layer-by-layer adsorption of charged polyelectrolytes are presented, too.

INTRODUCTION

Artificially grown diamond is a promising transducer material for chemical and biological sensing, as it is widely considered as biocompatible, displays outstanding electrical and electrochemical properties, and allows the direct coupling of biomolecules onto the diamond surface [1-5]. Among the various proposed concepts for the development of diamond-based chemical sensors and biosensors, the semiconductor field-effect platform is one of the most attractive approaches. Most of diamond-based field-effect (bio-)chemical sensors reported have been realized on a transistor structure by using hydrogen (H)-terminated polycrystalline or monocrystalline diamond films as an active transducer material [6-8].

Owing to the simplicity of the layout, the absence of a complicated encapsulation procedure and thus, an easier and cost-effective fabrication, capacitive field-effect structures are especially suited for (bio-)chemical sensor applications. Therefore, recently, we have introduced a capacitive field-effect electrolyte-diamond-insulator-semiconductor (EDIS) structure as a platform for (bio-)chemical sensing [9-11]. This work summarizes recent experimental results on the development of EDIS sensors for the detection of pH, penicillin concentration and layer-by-layer adsorbed charged macromolecules using an oxygen (O)-terminated nanocrystalline diamond (NCD) films as transducer material.

EXPERIMENTAL

Undoped NCD thin films of ~100 nm thickness were grown on a p-Si-SiO$_2$ (ρ=1-10 Ωcm, 50 nm thermally grown SiO$_2$) structure by means of microwave (2.45 GHz) plasma-enhanced chemical vapor deposition from a mixture of methane (CH$_4$) and hydrogen (H$_2$) in an ASTeX 6500 reactor. Prior to growth, the SiO$_2$ surface was seeded with a monodisperse colloid of nanocrystalline diamond particles in water with an ultrasonic bath. The NCD growth process on a SiO$_2$ as well as an additional surface treatment in oxidizing medium have been optimized to yield high pH-sensitive non-porous O-terminated NCD films without damage of the underlying SiO$_2$ layer. An Al film was deposited on the rear side of the Si chip as a contact layer. The chip size of the EDIS sensors has been 10 x 10 mm². For more details about the preparation of NCD films, see [12,13].

Typically, as prepared NCD surfaces are hydrogen (H)-terminated. In order to obtain O-terminated surfaces, the NCD films were treated in an oxidizing mixture of H$_2$SO$_4$ and KNO$_3$. The NCD films have been physically characterized by means of scanning electron microscopy (SEM), atomic force microscopy (AFM) and X-ray photoelectron spectroscopy (XPS) methods. The SEM micrograph in Figure 1 exemplarily demonstrates the surface morphology of a 100 nm thick NCD film. The film comprised randomly orientated fine grains and was totally closed. No visible pores in the film or damages of the underlying SiO$_2$ layer have been observed.

Figure 1. SEM picture of the surface morphology of a 100 nm thick NCD film grown on a p-Si-SiO$_2$ structure.

Additional AFM analysis of the diamond films was performed in intermittent contact mode (BioMat Workstation, JPK Instruments - Germany) and revealed a nanostructured surface consisting of NCD grains having a size of ~100 nm with an average surface roughness of 11 nm as depicted in Figure 2.

The EDIS sensors have been electrochemically characterized in buffer solutions with different pH values by means of capacitance-voltage and constant-capacitance (ConCap) method using an impedance analyzer (Zahner Elektrik). For the experiments, the EDIS sensors were mounted into a home-made measuring cell, sealed by an O-ring and contacted on its front side by the electrolyte and on its rear side by a gold-plated pin. A conventional liquid-junction Ag/AgCl electrode was used as a reference electrode. For more detailed information on the measurement set-up, see [9].

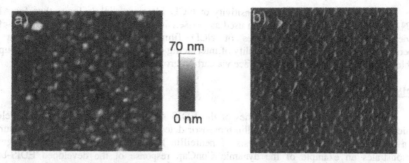

Figure 2. AFM height (a) and phase (b) images of a NCD film. The scan size is 2 x 2 μm².

RESULTS AND DISCUSSION

EDIS-based pH sensor

The pH sensitivity of O-terminated EDIS sensors has been studied by means of ConCap method. Figure 3 demonstrates exemplarily a typical dynamic pH-response (a) and calibration curve (b) of an EDIS structure with a 100 nm thick O-terminated NCD film oxidized by wet chemical oxidation for about 45 min. The ConCap signal has been recorded in Titrisol buffer solutions with different pH values from pH 4 to pH 12. The O-terminated EDIS sensors displayed a pH sensitivity of 48-50 mV/pH. The pH-sensitive properties of NCD films can be explained by the site-binding model, similar to ion-sensitive field-effect transistors. The surface potential of the diamond film could be influenced by the presence of hydroxyl groups at the oxidized diamond surface, resulting in the pH-dependent modulation of the space-charge capacitance in the semiconductor and the sensor output signal.

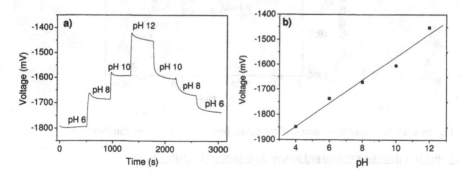

Figure 3. Typical ConCap response (a) and calibration curve (b) for O-terminated NCD film recorded in Titrisol buffer of different pH values from pH 6 to pH 12.

175

The obtained values of pH sensitivity of NCD films are slightly lower than for Al_2O_3, Si_3N_4, and Ta_2O_5 that have often been used as pH-sensitive transducer materials for field-effect devices [14]. The main advantages of NCD films are their chemical inertness and biocompatibility as well as the possibility of interfacing to biological systems by direct coupling of biomolecules onto the diamond surface via carbon groups.

EDIS-based penicillin biosensor

The high pH-sensitive properties of the EDIS structures have been used to develop a penicillin biosensor. The EDIS penicillin biosensor detects the local pH change near the surface as a result of the catalyzed hydrolysis of penicillin by the enzyme penicillinase. Figure 4 demonstrates an example of the dynamic ConCap response of the developed EDIS-based penicillin biosensor. In this experiment, the enzyme penicillinase (EC 3.5.2.6., *Bacillus cereus* from Sigma, specific activity: 1650 units/mg protein) was adsorptively immobilized directly onto the O-terminated NCD surface. The measurements have been performed in 0.5 mM polymix buffer solutions (pH 8, 100 mM KCl as an ionic-strength adjuster) with different penicillin concentrations from 5 µM to 2.5 mM. The penicillin solutions were prepared by dissolving penicillin G (benzyl penicillin, 1695 units/mg, Sigma) in the working buffer.

With increasing penicillin concentration, the concentration of the H^+ ions resulting from the penicillin hydrolysis is increased, too. As a result, the voltage that is necessary in order to adjust the constant capacitance raises. The freshly prepared NCD-based penicillin biosensor possessed a low detection limit of 5 µM and a high sensitivity of 65-70 mV/decade in the concentration range from 0.25 to 2.5 mM penicillin G.

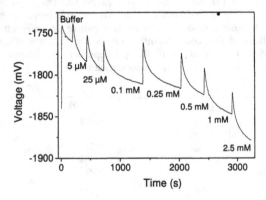

Figure 4. Typical ConCap response of the developed NCD-based penicillin biosensor.

Label-free detection of charged macromolecules with EDIS sensor

In order to demonstrate functional capabilities of NCD films for multi-sensing applications, the realized EDIS structures with O-terminated NCD film have been used for a label-free electrical detection of charged macromolecules. The positively charged PAH (Poly

176

(allylamine hydrochloride)) and negatively charged PSS (Poly (sodium 4-styrene sulfonate)) polyelectrolytes (PE) were chosen as model system. The multilayers of PAH/PSS were obtained by using the layer-by-layer assembly technique (see e.g., [15-17]) by sequential adsorption of PAH and PSS from the respective PE solution (50 µM PAH or PSS, pH 5.4). During the experiment, the EDIS sensors were consecutively exposed to the respective PE solution for a time necessary for the adsorption of each single monolayer, followed by rinsing in electrolyte solution. We started the formation of the PE multilayer onto O-terminated NCD with positively charged PAH.

Alternating potential changes, having the tendency to decrease with increasing the number of adsorbed polyelectrolyte layers (from 35-40 mV for first PE layers to 2-4 mV for an EDIS structure with 14-15 PE layers), have been observed after the adsorption of each PE layer. The adsorption of negatively charged PSS shifts the sensor signal towards the direction of a more negative surface charge, whereas the adsorption of the positively charged PAH shifts the sensor signal into the direction corresponding to a more positive charged NCD surface. Thus, the molecular layers induce an interfacial potential change resulting in alternating changes in the flat-band voltage, and therefore, in the output signal of the EDIS structure.

CONCLUSIONS

The presented results demonstrate the potential of field-effect capacitive EDIS structures having NCD films as transducer material for multi-parameter sensing of different (bio-)chemical quantities. The immobilization of other biomolecules, like DNA, proteins, on NCD films for extending the biosensor capabilities will be subject of future research.

ACKNOWLEDGMENTS

The authors thank H. P. Bochem for technical support. Part of this work was supported by the School for Life Sciences of the Transnational University Limburg, the IWT-SBO project "CVD diamond: a novel multifunctional material for high temperature electronics, high power/high frequency electronics and bioelectronics"), the Ministerium für Innovation, Wissenschaft, Forschung und Technologie des Landes Nordrhein-Westfalen, and the Bundesministerium für Bildung und Forschung (Germany).

REFERENCES

1. J. Rubio-Retama, J. Hernando, B. López-Ruiz, A. Härtl, D. Steinmüller, M. Stutzmann, E. López-Cabarcos, and J. A. Garrido, *Langmuir* **22**, 5837 (2006).
2. R. J. Hamers, J. E. Butler, T. Lasseter, B. M. Nichols, J. N. Russell Jr., K.-Y. Tse, and W. Yang, *Diamond Relat. Mater.* **14**, 661 (2005).
3. C. E. Nebel, B. Rezek, D. Shin, H. Uetsuka, and N. Yang, *J. Phys. D: Appl. Phys.* **40**, 6443 (2007).
4. V. Vermeeren, N. Bijnens, S. Wenmackers, M. Daenen, K. Haenen, O. A. Williams, M. Ameloot, M. van de Ven, P. Wagner, and M. Michiels, *Langmuir* **23**, 13193 (2007).
5. S. Wenmackers, V. Vermeeren, M. van de Ven, M. Ameloot, N. Bijnens, K. Haenen, L. Michiels, and P. Wagner, *Phys. Stat. Sol. (a)* **206**, 391 (2009).

6. K.-S. Song, M. Degawa, Y. Nakamura, H. Kanazawa, H. Umezawa, and H. Kawarada, *J. Appl. Phys.* **43**, L814 (2004).
7. K.-S. Song, G.-J. Zhang, Y. Nakamura, K. Furukawa, T. Hiraki, J.-H. Yang, T. Funatsu, I. Ohdomari, and H. Kawarada, *Phys. Rev. E* **74**, 041919 (2006).
8. C. E. Nebel, D. Shin, B. Rezek, N. Tokuda, H. Uetsuka, H. Watanabe, H., *J. R. Soc.Interface* **4**, 439 (2007).
9. M. H. Abouzar, A. Poghossian, A. Razavi, O. A. Williams, N. Bijnens, P. Wagner, and M. J. Schöning, *Biosens. Bioelectron.* **24**, 1298 (2009).
10. A. Poghossian, M. H. Abouzar, A. Razavi, M. Bäcker, N. Bijnens, O. A. Williams, K. Haenen, W. Moritz, P. Wagner, and M. J. Schöning, *Electrochim. Acta* **54**, 5981 (2009).
11. M. H. Abouzar, A. Poghossian, A. Razavi, A. Besmehn, N. Bijnens, O. A. Williams, K. Haenen, P. Wagner, and M. J. Schöning, *Phys. Stat. Sol. (a)* **205**, 2141 (2008).
12. O. A. Williams, O. Douhéret, M. Daenen, K. Haenen, E. Osawa, and M. Takahashi, *Chem. Phys. Lett.* **445**, 225 (2007).
13. M. Daenen, O. A. Williams, J. D'Haen, K. Haenen, and M. Nesladek, *Phys. Stat. Sol. (a)* **203**, 3005 (2006).
14. A. Poghossian, and M. J. Schöning, "Silicon-based Chemical and Biological Field-Effect Sensors", *Encyclopedia of Sensors,* vol. 9, ed. C. A. Grimes, E. C. Dickey, and M. V. Pishko (American Scientific Publishers, 2006) pp.1-71.
15. G. Decher, M. Eckle, J. Schmitt, and B. Struth, *Curr. Opin. Coll. Interface Sci.* **3**, 32 (1998).
16. M. Schönhoff, *Curr. Opin. Coll. Interface Sci.* **8**, 86 (2003).
17. M. Schönhoff, V. Ball, A. R. Bausch, C. Dejugnat, N. Delorme, K. Glimel, R. v. Klitzing, and R. Steitz, *Coll. Surf. A* **303**, 14 (2007).

Mater. Res. Soc. Symp. Proc. Vol. 1203 © 2010 Materials Research Society   1203-J08-05

Study of the Adhesion and Biocompatibility of Nanocrystalline Diamond (NCD) Films on 3C-SiC Substrates

Humberto Gomez[1, 3], Christopher L. Frewin[2], Ashok Kumar[1,4], Stephen E. Saddow[2,4], Chris Locke[2]

[1]Mechanical Engineering Department, University of South Florida, 4202 E. Fowler Ave, ENB 118, Tampa, FL, 33620 USA.
[2]Electrical Engineering Department, University of South Florida, 4202 E. Fowler Ave, ENB 118, Tampa, FL, 33620 USA.
[3] Departamento de Ingeniería Mecánica, Universidad del Norte, Barranquilla, Colombia.
4 Nanomaterials & Nanomanufacturing Research Center, University of South Florida, Tampa, FL 33620 USA.

ABSTRACT

The unique material characteristics of silicon carbide (SiC) and nanocrystalline diamond (NCD) present solutions to many problems in conventional MEMS applications and especially for biologically compatible devices. Both materials have a wide bandgap along with excellent optical, thermal and mechanical properties. Initial experiments were performed for NCD films grown on 3C-SiC using a microwave plasma chemical vapor deposition (MPCVD) reactor. It was observed from the atomic force microscopy (AFM) analysis that the NCD films on 3C-SiC possess a more uniform grain structure, with sizes ranging from approximately 5 – 10 nm, whereas on the Si surface, the NCD has large, non-uniform inclusions of grains ≈1 μm in size. The *in vitro* biocompatibility performance of NCD/3C-SiC was measured utilizing 2 immortalized neural cell lines: H4 human neuroglioma (ATCC #HTB-148) and PC12 rat pheochromocytoma (ATCC #CRL-1721). MTT (3-(4,5-Dimethylthiazol-2-yl)-2,5-diphenyltetrazolium bromide) assay was used to measure viability of the cells for 96 hours and live/ fixed cell. AFM was performed to determine the general cell morphology. The H4 cell line shows a good biocompatibility level with hydrogen treated NCD as compared with the cell treated polystyrene control well, while the PC12 cells show decreased viability on the NCD surfaces.

INTRODUCTION

The outstanding material characteristics of 3C-SiC and NCD present some unique solutions for the generation of high temperature, power, and high frequency MEMS devices. As semiconductor materials, 3C-SiC and NCD have very good electrical properties that can be utilized for the creation of electrical devices. The band gap difference and high carrier mobility for both of these materials generates a heterojunction that can become part of an extremely efficient and fast switching bipolar junction device [1]. As both of these materials are potentially biocompatible due to their chemical inertness, the generation of permanently implantable MEMS devices is also a possibility. Single crystal 3C-SiC is normally heteroepitaxially grown on silicon using CVD methods and the process is extremely sensitive to both growth and substrate conditions.

Alternatively, NCD can be deposited on any material stable at the deposition temperature by using microwave plasma chemical vapor deposition (MPCVD). However, NCD deposited on Si substrates possesses additional defects as the result of the stress originated by the lattice mismatch between them [2]. Consequently, the deposition of NCD on 3C-SiC substrates has been proposed in helping to alleviate this defects ($\alpha_{Si} = 0.543$ nm, $\alpha_{SiC} = 0.436$ nm, and $\alpha_{NCD} = 0.357$ nm) and generate a chemically inert surface with the potential use in a human body implantable device. In this paper, the deposition of NCD on 3C-SiC grown on (100) Si substrates is characterized in terms of the quality of the resulting NCD film. Additionally, the biocompatibility performance of the resulting NCD/3C-SiC / Si surface is compared with Si *in vitro* utilizing two immortalized cell lines, the H4 human neuroglioma and the PC12 rat pheochromocytoma cells.

EXPERIMENT

Nanocrystalline diamond was deposited on both 3C-SiC - which has been homoepitaxially grown on (100) Si -, and on Si (100) substrates. Prior to the diamond deposition, 3C-SiC and Si surfaces were seeded using a slurry solution of 100 ml of acetone containing diamond nanopowder and Ti powder at 1:1 ratio; samples were immersed on the slurry and left in an ultrasound bath for 20 minutes followed by two 20 minutes cleaning cycles in methanol. The samples are then removed and dried with N_2. AFM is used to characterize the resulting surfaces after the seeding process. A distribution of the diamond seeds is shown on Figure 1. From this analysis, it can be seen that more seeds appear to be bonded to the 3C-SiC surface compared to Si.

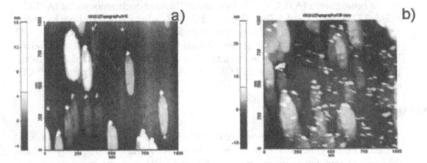

Figure 1. AFM images corresponding to the seeds distribution of, a) silicon, and b) 3C-SiC. Seeds can be distinguished as small rounded clumps, and solvent residues appear as enlarged- oval shaped features on the surface.

The diamond seeded substrates were placed on a MPCVD IPLAS *Cyrannus*® reactor. Intrinsic nanocrystalline diamond was grown using a pressure 135 Torr, 1.8 KW of microwave power, substrate temperature of 750°C, and a gas feedstock of 0.5% CH_4 / 98.5% Ar / 1% H_2 for a total flow of 800 sccm during 3 hours. Measurements using

Fourier transform infrared spectroscopy (FTIR) showed an uneven deposition with film thickness ranging from 400 – 600 nm. After the growth, NCD coated samples were exposed to hydrogen plasma in the same MPCVD system using 200 sccm of H_2, 2.5 KW of microwave power and 25 Torr of pressure. The purpose of this hydrogenation process is to obtain a hydrophobic positive charged surface with less sp_2 bonded carbon [3], and suitable for neurite/cell outgrowth formation and extension [4 - 5]. Further surface chemical treatments were performed prior to the subsequent biocompatibility tests to prevent bacterial growth and surface oxidation [5].

RESULTS AND DISCUSSION

Figure 2 shows AFM analysis of the surface morphology of NCD deposited on (100) Si, the initial surface of the 3C-SiC grown on (100) Si, and NCD deposited on the 3C-SiC / Si surface. NCD deposited on Si shows a regular, cubic grains ranging from 100 – 500 nm in perimeter and 1,000 nm^2 – 10,000 nm^2 in area. However, the surface is marked with craters and the presence of many large, oval shaped defects which measure approximately 5 µm in perimeter and 1µm^2 in area. The difference between these features on the NCD surface increased the roughness, R_q, to 17.0 nm, and might be the result of the non-uniform diamond seeds distribution on the surface showed in figure 1. The initial 3C-SiC, which was 3.45 µm thick measured by FTIR, shows the characteristic mesa growth pattern, atomic step terraces, and APB crevices. The initial surface has an Rq, of 7.25 nm. The NCD grown on the 3C-SiC / Si surface, with a more uniform diamond seeds distribution according to Figure 1, shows a relatively even-flat surface with many small cubic grains 75 - 300 nm in perimeter and 500 nm^2 – 7,500 nm^2 in area. This surface does not show the large grain inclusions displayed by the NCD on Si. The relatively flat surface and similar grain have surface roughness R_q of 7.30 nm, which is a little over half the roughness of the NCD on Si.

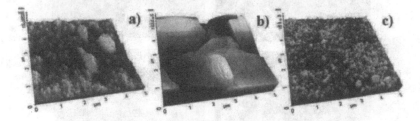

Figure 2. Three dimensional AFM micrographs of a) NCD deposited on (100) Si, b) 3C-SiC grown on (100) Si, and c) NCD deposited on 3C-SiC / Si.

Figure 3 shows the corresponding Raman spectrum of the NCD films grown on Si and 3C-SiC /Si. Typical spectrum features of NCD film grown in presence of 1% hydrogen in the gas chemistry were observed for Si and 3C-SiC /Si substrates, the 1140 cm^{-1} indicate the trans-polyacetylene state. A small peak shift is observed for the spectrums. Figure 4 shows further examination of the morphology of the samples through the use of scanning electron microscope. NCD on 3C-SiC /Si, shows a relatively smooth surface in 4-a), which has small grain and very small grain defects. The cross section in 4-b) reveals that the film is highly granular, possessing many random grain boundaries. NCD deposited on Si shows a surface which possesses small grains with intermixed larger grains, much like was seen in the AFM results. A difference in this film when compared to the NCD on 3C-SiC /Si film is an evidence of a not well adherent NCD/ Si interface, probably because the internal stresses generated during the growth due the lattice mismatch; as shown in 4-d) the film is broken back and uneven as compared to the 3C-SiC /Si which is even with the cleaved surface.

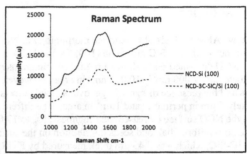

Figure 3. Raman spectra of 1% H_2 NCD films on Si and 3C-SiC / Si.

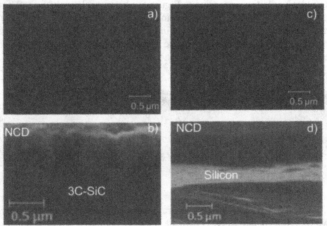

Figure 4. SEM plan view and cross-section micrographs of the NCD deposited on 3C-SiC / Si and Si. NCD on 3C-SiC is displayed in (a-b), and on Si in (c-d)

NCD Biocompatibility performance

Hydrogen terminated NCD was tested with 2 immortalized cell lines: H4 human neuroglioma (ATCC #HTB-148), and PC12 Rat pheochromocytoma (ATCC #CRL-1721). Initial biocompatibility testing was performed by culturing these two cell lines on the substrate *in vitro*. MTT (3-(4, 5-Dimethylthiazol-2-yl)-2,5-diphenyltetrazolium bromide) assay was used to measure viability of the cells for 96 hours and live/ fixed cell. In this assay, MTT is metabolized by living mitochondria to produce purple formazan which is soluble in dimethyl sulfoxide and subsequently quantified by using a spectrophotometer measuring a wavelength of 500 to 600 nm [4]. Figure 5-a) shows a bar graph of the results of the MTT and indicates that the H4 cell line possessed good viability with NCD when compared with the cell treated polystyrene control, while NCD shows decreased viability with the PC12 cell line. Figure 5-b) uses AFM analysis to show that neuronal cells appear to metabolize or react chemically with Si surface, whereas NCD surfaces are unaffected by cells. Intrinsic NCD showed a similar performance profile with the H4 cells as Si (100) by possessing a good level of viability. Lamellipodia and filopodia extensions of the cell, which are extensions that are used for motility and forming networks with other cells, showed a good level of cellular membrane expansion and attachment to this substrate through the use of AFM techniques [5]. Alternatively, there was a poor level of viability with the PC12 line and the high profile of the cell soma shows it does not favor this surface. Hydrogen treated NCD has been shown by Ostrovskaya et al. (via contact angle measurements of 75° - 95°) to be a hydrophobic surface [6]. Fine et al. demonstrated that neurite formation and extension is favorable for positive charged surfaces [7]. H4 cell behavior seems to be in accordance with these previous studies but the low level of lamellipodia attachment for PC12s may be due to the intrinsic nature of the bulk substrate and a non uniform distributed positive charge surface. Further exploration into the surface properties of NCD and how this material interacts with extracellular proteins are warranted from these initial results.

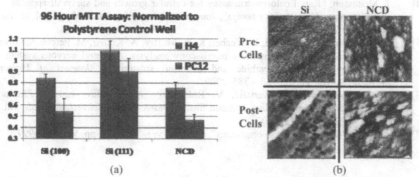

(a)   (b)

Figure 5. (a) Displays a bar graph representation of the MTT assay results for the H4 and PC12 cell lines. The graph is expressed as the s.d.m., $\bar{x}$, with error bars indicating the s.e.m., σ_{x_m}, from the three MTT assays performed on triplicate samples and normalized to the cell treated polystyrene control well. (b) Displays the 5 μm x 5μm AFM micrographs of Si (100) and NCD surfaces before and after cell seeding. The Si shows a pot marked and damaged surface whereas the NCD surface appears morphologically similar..

CONCLUSIONS

NCD was successfully deposited on both Si and 3C-SiC surfaces using plasma MWPCVD. NCD grown on 3C-SiC displayed superior morphology compared to NCD grown on Si. The grains and defective clusters are smaller with the NCD deposited on 3C-SiC than that of the NCD deposited on Si. The NCD morphology on 3C-SiC is smother and independent of 3C-SiC texture. *In* vitro biocomptatbility shows that hydrogen treated diamond had some good viability and cell attachment with The H4 cell lines, but the hydrophobic and a possible non uniform positive charge of surface was not as suitable for the PC12 cell line. Further investigation into the biocompatability mechanisms must be performed to explore the biocompatibility of NCD further.

REFERENES

[1]   K. Das, "Diamond and silicon carbide heterojunction bipolar transistor", U. S. Patent 5,285,089, February 8, 1994.

[2]   H. Yoshikawaa, C. Morela, and Y. Kogab, "Synthesis of nanocrystalline diamond films using microwave plasma CVD", *Diamond and Related Materials*, vol. 10, no. 9 – 10, , pp. 1588 – 1591, 2001.

[3]   S.Jeedigunta, Z.Xu, M.Hirai, P.Spagnol, A.Kumar, *"Effects of plasma treatments on the nitrogen incorporated nanocrystalline diamond films"*, Diamond & Related Materials, 17 (2008), 1994 – 1997.

[4]   T. Mosmann, "Rapid colorimetric assay for cellular growth and survival: application to proliferation and cytotoxicity assays", *Journal of immunological methods*, vol. 65, no 1 – 2, pp. 55 – 63, 1983.

[5]   C. L. Frewin, M. Jaroszeski, E. Weeber, K. E. Muffly, A. Kumar, M. Peters, A. Oliveros and S. E. Saddow, "Atomic force microscopy analysis of central nervous system cell morphology on silicon carbide and diamond substrates", *Journal of Molecular Recognition*, vol. 22. pp. 380 – 388, 2009.

[6]   L. Ostrovskaya, V. Perevertailo, V. Ralchenko, A. Saveliev, and V. Zhuravlev, "Wettability of nanocrystalline diamond films", *Diamond and Related Materials,* vol. 16, no. 12, pp. 2109 – 2113, 2007.

[7]   A. Rouhi, "Contemporary biomaterials", *C & EN,* vol. 77, no. 3, pp. 51 – 59, 1999.

Mater. Res. Soc. Symp. Proc. Vol. 1203 © 2010 Materials Research Society 1203-J15-03

MESFETs on H-terminated Single Crystal Diamond

P. Calvani[1], M.C. Rossi[1], G. Conte[1], S. Carta[1,2], E. Giovine[2], B. Pasciuto[3], E. Limiti[3], F. Cappelluti[4], V. Ralchenko[5], A. Bolshakov[5], G. Sharonov[6]

[1] Electronic Engineering Dept., Roma Tre University, Roma, Italy
[2] IFN-CNR, Roma, Italy
[3] Electronic Engineering Dept., Tor Vérgata University, Roma, Italy
[4] Department of Electronics, Politecnico di Torino, Torino, Italy
[5] General Physics Institute, Russian Academy of Science, Moscow, Russia
[6] Institute of Applied Physical Problems, Belarus State University, Minsk, Belarus

ABSTRACT

Epitaxial diamond films were deposited on polished single crystal Ib type HPHT diamond plates of (100) orientation by microwave CVD. The epilayers were used for the fabrication of surface channel MESFET structures having sub-micrometer gate length in the range 200-800 nm. Realized devices show maximum drain current and trasconductance values of about 190 mA/mm and 80 mS/mm, respectively, for MESFETs having 200 nm gate length. RF performance evaluation gave cut off frequency of about 14 GHz and maximum oscillation frequency of more than 26 GHz for the same device geometry.

INTRODUCTION

Current semiconducting materials do not offer high power RF (> 8 GHz) devices in simple solid state device configurations, required for compact MMIC usually employed for communication and radar applications. Among wide band gap materials, diamond has by far the optimum material characteristics allowing for, at least in principle, the best power amplification per unit gate length at microwave frequencies, to be employed in the fields of electrical power management and wireless communications.

Owing to the encouraging high frequency and power performance, the technology of hydrogenated diamond is currently receiving much interest for the fabrication of surface channel metal semiconductor field effect transistors (MESFETs) [1-3]. In this structure, hydrogen termination of diamond together with carrier transfer into surface acceptor states promotes an upward surface band bending which gives rise to a space charge extending several nanometers below the surface without the addiction of extrinsic doping impurities. Here holes are confined perpendicularly to the surface from the electrostatic field arising from the charge separation (positive charge in diamond and negative on the surface acceptors) whereas are free to move in parallel to the surface [4], where they form a 2DHG closely following the surface topography. In this way, surface hole densities up to 10^{13} cm^{-2} and mobility values about 100-200 cm^2/Vs cm^{-2} are typically achieved [5].

In this context, two different strategies are currently pursued for surface channel device realization, based on polycrystalline and on single diamond substrates. In the former case large substrates required for electronic applications are already available, although achieved results still appear to be affected by the polycrystalline sample structure. At variance, the realization of diamond electronics on epitaxial layer appears to be limited to small areas (few squared mm), although encouraging results both in terms of high frequency and power density can be found in literature. In this paper we present results achieved for devices fabricated on homoepitaxial diamond layers grown on HPHT diamond substrates by Microwave Plasma Enhanced Chemical Vapor Deposition (MPCVD).

EXPERIMENTAL

Butterfly shaped Metal Semiconductor Field Effect Transistors (MESFETs) have been fabricated on Single Crystal Diamonds. Homoepitaxial diamond layers of 0.5 and 1.0 μm thickness have been grown by MPCVD in CH_4 (2%)-H_2 gas mixtures on (100) oriented HPHT diamond plates polished to the surface roughness R_a ~4 nm. The deposition has been terminated in a microwave hydrogen plasma (for 10 min) in order to induce typical band bending, responsible for the generation of a p-type conductive surface channel with sheet resistivity of ~20 kΩ/□. Ohmic drain and source contacts have been realized with gold evaporation on 1 μm thick and 4.0 X 4.0 um^2 area films while aluminum was used for self-aligned gate Schottky contact. Electron Beam Lithography (EBL) has been used for the realization of gate length from 800 nm down to 200 nm. Adopted device scheme is sketched in figure 1.
Following a DC characterization performed by using a probe station equipped with a HP4140B picoammeter, bias dependent S-parameter measurements of devices under test have been performed in the 50MHz÷20GHz frequency range. S-parameters were measured by a VNA calibrated with *off-wafer SOLT* procedure. The measurement set-up consists in: HP 8510C Vector Network Analyzer, Cascade RF1 Probe Station, Coplanar Probe GGB Picoprobe (pitch=200μm), CS5 Calibration kit.

RESULTS AND DISCUSSION

Figure 2 shows the DC-output characteristics of a surface-channel-MESFET with 200 nm gate length Lg and 25 μm gate width W, in the gate bias voltage range -2.5-1 V. A maximum drain to source current of 180 mA/mm has been obtained. As shown, short gate length devices exhibit a non complete channel closure, even when the gate junction is slightly forward biased. Such a behavior is tentatively related to short channel effects, although the existence of a residual conductive channel uncontrolled by the gate electrode cannot be ruled out.

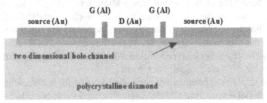

Figure 1. Device scheme cross section.

DC input characteristics of the same device are reported in figure 3. At variance with the usually found square law dependence of I_{DS} on V_{GS}, indeed detected for devices with 400 and 800 gate length, an almost linear trend, usually observed in short channel devices, is detected for Lg=200 nm, when the device is operating in the saturation region, with $V_{DS} \geq 4$ V. For

$V_{DS} < 4$ V a linear trend is still observed, although exhibiting a smaller slope and holding in a different V_{GS} interval. Furthermore, a small but not negligible drain to source current is detected for low V_{GS} and related to sub-threshold conduction processes. The onset of short channel effects can be also evidenced from the equally spaced output current-voltage characteristics and is further corroborated by a threshold voltage V_T shifting from 0 to 0.5 V, as detected when the gate length is changed from 800 to 200 nm (not shown).

Since a negative gate to source voltage is able to modulate the channel current, a quasi-enhancement behavior has been attributed to the diamond FET. Field effect holes mobility extracted from input characteristics appears field dependent at low V_{DS}, approaching a constant value of about 150 cm^2/Vs for $V_{DS} \geq 4$ V, when a constant slope of the trans-characteristics is found. Such a field effect mobility value is also confirmed by Hall measurements performed on un-gated hall bar structures. The corresponding trans-conductance is reported in figure 4. The maximum value reached for V_{DS}=-10.0 V was more than 80 mS/mm. As usually reported in hetero-structure devices, a transconductance peak slightly shifting toward high V_{GS} is found for increasing V_{DS}.

The measured frequency dependence of the current gain $|H_{21}|^2$ and Maximum Available Gain (MAG) obtained at V_{DS}=-10 V and V_{GS}=-0.3 V are reported in figure 5 for the same device with L_g=0.2 μm and W=25 μm. Device show a f_{MAX} of more than 26 GHz with a gain of 22 dB at 1 GHz while the achieved threshold frequency f_T was about 14 GHz.

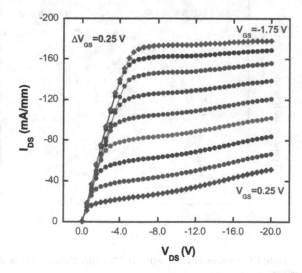

Figure 2. Current voltage output characteristics for a MESFET with 200 nm gate length.

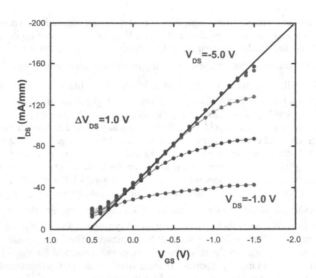

Figure 3. Current-voltage input characteristics for a MESFET with 200 nm gate length. Continuous lines are only guidelines.

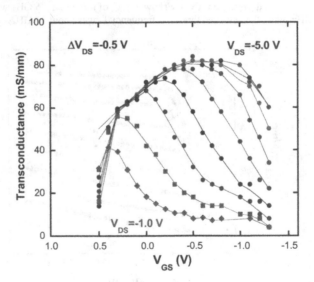

Figure 4. Typical MESFET transconductance with MESFET with 200 nm gate length. Continuous lines are only guidelines.

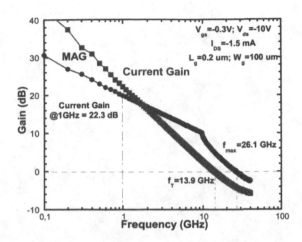

Figure 5. Current gain and Maximum Available Gain (MAG) for a MESFET with 200 nm gate length and gate width W=25 μm.

CONCLUSIONS

Epitaxial diamond films have been grown by MPCVD on polished single crystal Ib type HPHT diamond plates of (100) orientation. Such epilayers have been employed for the fabrication of surface channel MESFET structures having sub-micrometer gate length (200-800 nm). A detailed DC and RF characterization has been carried out on realized devices, giving maximum I_{DS} and trasconductance values of about 190 mA/mm and 80 mS/mm, respectively, for MESFETs having 200 nm gate length. RF performance evaluation have been carried out through S parameters analysis, leading to $f_T \sim 14$ GHz and f_{max} 26 ~GHz for the same device geometry. Such values, although not yet optimized, confirmed the achieved high quality of epitaxial diamond film and hydrogen termination.

REFERENCES

1. H. Taniuchi, H. Umezawa, T. Arima, M. Tachiki, H. Kawarada, *IEEE Electron Dev. Lett.* **22,** 390 (2001).
2. M. Kasu, K. Ueda, H. Ye, Y. Yamauchi, S. Sasaki, T. Makimoto, *Diamond Relat. Mater.* **15,** 783 (2006).
3. H. Ye, M. Kasu, K. Ueda, H. Ye, Y. Yamauchi, N. Maeda, S. Sasaki, T. Makimoto, *Diamond Relat. Mater.* **15,** 787 (2006).
4. F. Maier, J. Ristein, L. Ley, "Origin of the surface conductivity in diamond", *Phys. Rev. B* **85,** 3472 (2000).
5. D. Takeuci, M. Riedel, J. Ristein, L. Ley, *Phys. Rev. B* **68,** 4130 (2003).

Device Applications

(2) Electrochemistry

Mater. Res. Soc. Symp. Proc. Vol. 1203 © 2010 Materials Research Society   1203-J07-01

Electrochemical Charge Transfer to Diamond and Other Materials

Vidhya Chakrapani[1], John C. Angus[1,a], Kathleen Kash[2], Alfred B. Anderson[3]
[1]Chemical Engineering Department, [2]Department of Physics, [3]Department of Chemistry,
Case Western Reserve University, Cleveland, OH 44106, U.S.A.

Sharvil Desai[4] and Gamini U. Sumanasekera[4]
[4]Department of Physics, University of Louisville, Louisville, KY 40292, U.S.A.

[a] corresponding author

ABSTRACT

The oxygen redox couple in adsorbed water films acts as an "electrochemical ground" that tends to pin the Fermi level in solids at the electrochemical potential of the redox couple. We discuss this effect on the conductivity of diamond; the conductivity type of sp^2-based carbons including single-walled, semiconducting carbon nanotubes and graphene; the photoluminescence of GaN and ZnO; and the contact charging of metals.

INTRODUCTION

Early studies

In 1989 Maurice Landstrass and K.V. Ravi reported that hydrogen-terminated diamond showed a p-type surface conductivity when exposed to air [1]. This observation was extremely unusual because no similar observation had been reported in more than a century of studying the properties of diamond. Also unusual was the observation that the surface conductivity is not seen in diamond synthesized at high pressure and high temperature. This set of highly unusual characteristics has made the surface conductivity a subject of much discussion and its mechanism is still somewhat controversial.

The first studies that led to eventual understanding of the effect were performed by Gi and co-workers [2-4]. They showed that the sheet hole concentration could be as great as 10^{13} cm^{-2}, and that the conductivity increased on exposure to acidic gases and decreased with exposure to basic gases. They proposed that the effect arose from surface oxidation of the diamond by the hydronium ion, H_3O^+. Subsequently, Maier et al. [5] described this oxidation process in terms of the hydrogen redox couple, $2H^+ + 2e^- = H_2$, operating in an adsorbed water film. Later studies by Foord [6] et al. and Chakrapani [7] et al. indicated that the oxygen redox couple was responsible for the effect. Chakrapani confirmed this conclusion by demonstrating that the changes in pH and dissolved oxygen concentration upon adding diamond powder to macroscopic size aqueous solutions were consistent with the four-electron oxygen redox couple, $O_2 + 4H^+ + 4e^- = 2H_2O$ [8]. It is of interest that electron transfer to aqueous redox couples is not the first report of surface transfer doping in diamond. Shapoval et al. of the Institute of General and Inorganic Chemistry in Kiev in 1995 invoked electron transfer to an electrochemical couple to explain the surface conductivity of diamond immersed in molten salts [9].

Electrochemically mediated surface transfer doping of diamond to an aqueous redox couple is a particular example of a more general process [10]. Other adsorbed acceptors have been used. For example, Strobel *et al.* [11] used an adsorbed layer of C_{60} molecules and Qi and co-workers [12] used tetrafluorotetracyanoquino-dimethane to induce p-type conductivity in diamond. Furthermore, transfer doping occurs in other semiconductor systems. Chakrapani and co-workers showed its effect on the luminescence of both GaN and ZnO [13]. Recently, Sque *et al.* presented theoretical arguments showing that it should be possible to use the oxygen redox couple for the transfer doping of graphene [14]. Chen and co-workers have recently published an extensive review on surface transfer doping of semiconductors [15]. Related effects in other systems have been discussed by Liu and Bard [16].

Electrochemistry and physics

Electrochemical transfer doping is truly an interdisciplinary field that has been approached both by electrochemists and by solid-state physicists. These fields each bring their own nomenclature to describe the same phenomena. Table I is a "translation table" relating terms commonly used in solid-state physics and electrochemistry.

Table I. Translation Table for Physicists and Electrochemists

Physicists	Electrochemists
Fermi energy	Electrochemical potential of an electron
Acceptor	Oxidizing agent
Donor	Reducing agent
Reference *energy* is electron at rest in vacuum.	Reference *potential* is the potential of the standard hydrogen electrode.

The relationship between the electrochemical and physical scales is given by equation 1.

$$\varepsilon = C + (-1)E \qquad\qquad (1)$$

where ε is the electron energy in electron volts and E is the potential in volts versus the standard hydrogen electrode. The offset between the two scales, C, is not known exactly and there is a residual uncertainty of as much as 0.1 eV [17-19]. The relationship between the potential and energy scales is shown in figure 1.

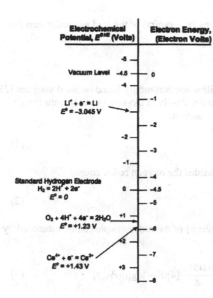

Electrochemical Potential, E^{SHE} (Volts)	Electron Energy, ε (Electron Volts)
-5	
Vacuum Level -4.5	0
-4	
	-1
Li$^+$ + e$^-$ = Li	
E^o = -3.045 V	
	-2
-2	
	-3
-1	
	-4
Standard Hydrogen Electrode	
H$_2$ = 2H$^+$ + 2e$^-$ 0	-4.5
E^o = 0	-5
O$_2$ + 4H$^+$ + 4e$^-$ = 2H$_2$O +1	
E^o = +1.23 V	-6
+2	
	-7
Ce^{4+} + e$^-$ = Ce^{3+}	
E^o = +1.43 V +3	
	-8

Figure 1. Relationship between the electrochemical potential scale and the physical scale of electron energies. The reference state for the electrochemical scale is the potential of the standard hydrogen electrode; the reference state for the physical scale is the electron at rest in vacuum.

THEORY

Electron electrochemical potential and the Fermi energy

The Fermi energy (Fermi level) and the electrochemical potential of an electron are identical concepts [20]. When equilibrium is achieved between metals and semiconductors in intimate contact, the Fermi energy (electrochemical potential of the electrons) is the same in all of the phases.

Fixing the Fermi energy at some constant value is commonly called "pinning the Fermi level." Two requirements are necessary for Fermi level pinning: a "large" electron reservoir and "rapid" electron exchange between the system and the reservoir. If either one is absent, the Fermi level of the system will not be pinned. In this context "rapid" means that the time constant for electron exchange must be much less than the time constant for changes that can occur in the system. The reservoir must be large enough so the properties of the reservoir are not changed significantly by the electron transfer to or from the system. If these conditions are met, in equilibrium, the system is "pinned" at the Fermi energy of the reservoir. (In statistical mechanics the limiting case of this concept is the grand canonical ensemble.) For a semiconductor exposed to humid air there are essentially infinite reservoirs of oxygen, water and carbon dioxide, which are required for the redox couple (see below). However, the requirement of rapid electron exchange may not always be met.

In chemistry an analogous situation to Fermi level pinning is the use of buffer solutions to fix the pH. In this case, a large excess of an acid, e.g., acetic acid, and the corresponding anion, e.g., acetate ion, tend to maintain a constant pH. The resistance to change of the solution pH depends on the presence of large concentrations (reservoirs) of the acid and anion and the relatively rapid kinetics of ionic reactions.

Electrochemistry in adsorbed water films

It is well known that in humid air water films are commonly found on solid surfaces [21]. It is less well recognized that the water films contain dissolved oxygen and CO_2 from the air. The dissolved CO_2 generates acidity through the reaction,

$$CO_2 + H_2O = HCO_3^- + H^+ \tag{1}$$

The presence of dissolved oxygen and protons enables the oxygen redox couple.

$$O_2 + 4H^+ + 4e^- = 2H_2O \tag{2}$$

The electron electrochemical potential (Fermi energy) of the redox couple can be estimated by applying the Nernst equation to reaction (2).

$$\mu_e(eV) = -4.44 + (-1)(+1.229) + \frac{0.0592}{4}[4pH - \log_{10}(P_{O_2})] \tag{3}$$

The overall chemical reaction that takes place during electrochemical transfer doping of diamond can be written [7],

$$4\left(e^-h^+\right)_{dia} + O_{2,air} + 2H_2O_{air} + 4CO_{2,air} = 4h_{dia}^+ + 4HCO_{3\,film}^- \tag{4}$$

In the forward direction of reaction 4 the diamond is oxidized, producing holes in the valence band and charge-compensating bicarbonate anions in the adsorbed water film. In the reverse direction, holes are consumed and oxygen is regenerated. Electrons will transfer between the redox couple and the semiconductor in a direction that brings the Fermi levels into alignment.

The type of compensating anion depends on the details of the experimental situation. In air, the dissolved CO_2 will fix the pH close to 6, and the dominant anions are bicarbonate ions, HCO_3^-. At higher pH, carbonate ions would also be present. If HCl vapor is used to adjust the pH to low values, the chloride ion will dominate. These charge-compensating anions are likely hydrated and adsorbed close to the diamond surface.

The relative positions of the oxygen redox couple at pH = 0 and pH = 14 and the band edges of several common semiconductors are shown in figure 2. Note that for diamond the redox couple spans the valence band maximum; for semiconducting, single-walled carbon nanotubes the oxygen redox couple essentially spans the entire band gap; and for GaN and ZnO the redox couple at all pH is within the band gaps. These different placements give rise to very different phenomena that are discussed in the following sections.

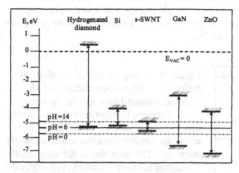

Figure 2. Relationship of the band structure of several semiconductors with the oxygen redox couple at pH of 0 and pH of 14. Figure taken from Chakrapani *et al.* [8].

Caveats

Although Fermi level pinning by the oxygen redox couple appears to be a common phenomenon, several important caveats must be emphasized.

1. The oxygen redox reaction is slow unless catalyzed. Therefore, the electron exchange between the semiconductor and the environment may not always be sufficiently rapid to pin the Fermi level.

2. The evidence for electron transfer between a macroscopic water phase and diamond is unequivocal [8]; however, the evidence for adsorbed water films on diamond [22] and on many semiconductors is indirect. The presence of water films may depend on surface defects and/or impurities. Furthermore, the thermodynamic properties of thin water films can be different than the properties of bulk water.

3. Other chemical processes, such as chemical ionization of surface functional groups that do not involve electron transfer, can occur in parallel with the electrochemical effect.

4. Organic films arising from glues, photo masks or insulation can mask the effect.

5. The electron affinities and work functions of solids in contact with water are different than those measured in ultra high vacuum.

6. There is a residual uncertainty in the value of the offset, C, between the electrochemical potential scale and the physical scale of electron energy [19].

OBSERVATIONS

Diamond

The p-type conductivity in hydrogen-terminated diamond arises when the electron electrochemical potential of the redox couple is below the Fermi level of the diamond. Electrons transfer to the redox couple, reducing hydronium ions and oxygen to water according to the forward direction of reactions 2 and 4. A space charge layer of holes forms in the diamond, protons are consumed and charge-compensating anions are generated in the water film. At high pH, the electron electrochemical potential of the redox couple is increased and the electron transfer is in the reverse direction, quenching holes and increasing the acidity. These processes are shown in figure 3 taken from Chakrapani *et al.* [8].

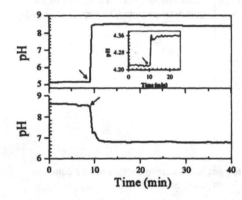

Figure 3. Change in pH of solutions of different acidities upon addition of air-exposed diamond powder. In the upper panel electrons transfer from the diamond into the acidic solution; in the lower panel electrons transfer from the solution into the diamond. The inset shows the very small effect that is observed when chemically de-oxygenated water is used. Figure taken from Chakrapani *et al.* [8].

The surface conductivity was not observed prior to 1989 because of the common protocol for preparing diamond for electrical measurements. If any surface conductivity was observed, it was attributed to surface impurities and the diamonds were heated in oxidizing acid. This procedure left the diamond with an oxygen-terminated surface. The resultant surface dipole, opposite in sense to that of a hydrogen-terminated surface, lowered the electron energies in the diamond to the point where the valence band edge was below the oxygen redox couple at all pH. From 1989 on, it was common to clean diamonds with atomic hydrogen, which left the surface hydrogen-terminated.

The surface conductivity is not observed with synthetic diamonds grown at high temperature and high pressure because of the ubiquitous presence of non-aggregated nitrogen in these crystals. Substitutional nitrogen, while a deep donor to the conduction band, can serve as a donor to quench holes formed in the valence band [23].

sp^2-based carbon systems

The sp^2-based carbons share a common feature: their work functions are similar and sufficiently small that the Fermi levels are all above the electron chemical potential of the oxygen redox couple at pH ~ 6. (The work functions are shown on the left hand side of figure 4.) Accordingly, all of these carbons can transfer electrons to the redox couple when exposed to humid air, which is illustrated by the downward pointing arrow.

The electron affinities of some common highly oxidizing gases are shown on the right hand side of figure 4. It is clear from the figure that there is a significant energy penalty for transfer of electrons from the sp^2 carbons to these dry gases (upward pointing arrow). The presence of water greatly changes the energetics of the reaction and makes electron transfer to the oxygen redox couple possible.

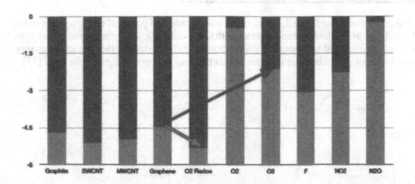

Figure 4. Work functions (in electron volts) of several sp²-based carbons (left hand side) and electron affinities of several common oxidizing gases (right hand side) compared to the Fermi energy of the oxygen redox couple at pH ~ 6 (middle of figure). Note that electrons can transfer from the sp² carbons into the redox couple, but that there is a large energy penalty for electron transfer into the oxidizing gases. Note: SWCNT refers to single-walled carbon nanotubes; MWCNT refers to multi-walled carbon nanotubes.

In figure 5 the change in Seebeck coefficient of vacuum-annealed, multi-walled carbon nanotubes from n-type to p-type upon exposure to air is shown. This change is in agreement with the above discussion. Similar changes, not shown here, in agreement with figure 4 and the predictions of Sque *et al.* [13], are observed with graphene [24]. From figure 2 it is clear that the energy range of the oxygen redox couple essentially spans the band gap of single-walled, semiconducting nanotubes. Therefore, pinning the Fermi level by the couple at high pH should lead to n-type conductivity; pinning at low pH should lead to p-type conductivity. These expectations are also in accord with observations [24, 25].

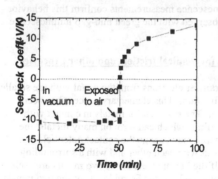

Figure 5. Change in Seebeck coefficient of vacuum-annealed multiwalled carbon nanotubes upon exposure to air. Data are from G. Sumanasekeran and S. Desai [24].

Gallium nitride, zinc oxide and other semiconductors

Again referring to figure 2, the electron electrochemical potential of the oxygen redox couple at all pH lies within the band gaps of both GaN and ZnO. This means that the oxygen redox couple can be used to change the occupation of mid-gap states if they are in the energy range of the couple. A schematic diagram of this process is shown in figure 6 for GaN [13].

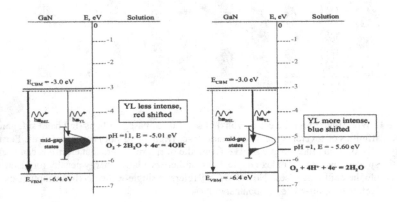

Figure 6. Luminescent channels in GaN upon exposure to NH_3 vapors at pH = 11 (left hand panel) and exposure to HCl vapors at pH = 1 (right hand panel). Note: YL refers to yellow luminescence. Figure taken from Chakrapani *et al.* [13].

At high pH, the Fermi level will tend to fill the midgap states associated with the yellow, luminescence; at low pH these states are partially emptied. Since the band edge luminescence and the mid-gap luminescence are in competitive channels, the yellow luminescence is diminished at high pH and enhanced at low pH; the ultra-violet luminescence from the band edge transition moves in the opposite direction. Luminescence measurements confirm this behavior [13]. Related behavior, differing in detail, was observed with the green mid-gap luminescence from ZnO.

Speculations: contact electrification of metals, mechanical friction, and other processes

Two metals when placed in contact will transfer electrons from the metal with the smaller work function to the metal with the larger work function. This elementary process has been confirmed many times for clean metals in vacuum. The situation for metals in contact with humid air is much different. In addition to oxide films, which can form on many metals, the presence of a water film and the oxygen redox couple may change the contact potential. For example, consider two clean metals in direct contact with each other and with a surrounding water film containing the oxygen redox couple. If the Fermi levels of the two metals are both above that of the redox couple, both will charge positively; if the Fermi levels of the two metals are below that of the oxygen redox couple, both will charge negatively. In vacuum the two metals would have opposite charge. One can speculate that the resultant electrostatic forces will influence important physical processes, for example, contact electrification of particulates and mechanical friction. Furthermore, we believe that similar processes may have evolved in biological processes such as insect locomotion.

CONCLUSIONS

Fermi level pinning by the oxygen redox couple in adsorbed water films is a ubiquitous process that has not been widely recognized. It can influence measurements of electrical and optical properties of semiconductors exposed to humid air. It must be taken into account when analyzing the behavior of semiconductors that operate in humid air or water, for example, sensors. Furthermore, it may play an unsuspected role in other physical processes such as contact electrification and mechanical friction.

ACKNOWLEDGMENTS

The authors gratefully acknowledge the support of the National Science Foundation: Division of Materials Science, grant DMR-9901419; Division of Chemistry, grants CHE-0314688 and CHE-0809209; and the Division of Electrical, Communications and Cyber Systems, grant ECCS-0925835. Useful discussions were held with Hiroshi Kawarada and Mohan Sankaran.

REFERENCES

1. M. I. Landstrass and K. V. Ravi, *Appl. Phys. Lett.* **55**, 975 (1989).
2. R. S. Gi, T. Mizumasa, Y. Akiba, Y. Hirose, T. Kurosu and M. Iida, *Jap. J. Appl. Phys. Part 1*, **34**, 5550 (1995).
3. R. S. Gi, T. Ishikawa, S. Tanaka, T. Kimura, Y. Akiba and M. Iida, *Jap. J. Appl. Phys. Part 1*, **36**, 2057 (1997).
4. R. S. Gi, K. Tashiro, S. Tanaka, T. Fujisawa, H. Kimura, T. Kurosu and M. Iida, *Jpn. J. Appl. Phys.* **38**, 3492 (1999).
5. F. Maier, M. Riedel, B. Mantel, J. Ristein and L. Ley, *Phys. Rev. Lett.* **85**, 3472 (2000).
6. J. S. Foord, C.H. Lau, M. Hiramatsu, R.B. Jackman, C.E. Nebel and P. Bergonzo, *Diamond and Related Mat.* **11**, 856 (2002).
7. V. Chakrapani, S. C. Eaton, A. B. Anderson, M. Tabib-Azar and J. C. Angus, *Electrochem. and Solid State Lett.* **8**, E4 (2005).
8. V. Chakrapani, J. C. Angus, A. B. Anderson, S. D. Wolter, B. R. Stoner and G. U. Sumanasekera, *Science* **318**, 1424 (2007).
9. V. I. Shapoval, I. A. Novosyolova, V. V. Malyshev and H. B. Kushkhov, *Electrochim. Acta* **40**, 1031 (1995).
10. J. Ristein, *Science* **313**, 1057 (2006).
11. P. Strobel, M. Riedel, J. Ristein and L. Ley, *Nature* **420**, 439 (2004).
12. D. Qi, W. Chen, X. Gao, L. Wang, S. Chen, K.P. Loh, and A. T. S. Wee, *J. Am. Chem. Soc.* **129**, 8084 (2007).

13. V. Chakrapani, C. Pendyala, K. Kash, A. B. Anderson, M. K. Sunkara and J. C. Angus, *J. Am. Chem. Soc.* **130**, 12944 (2008).
14. S. J. Sque, R. Jones and P. R. Briddon, *phys. stat. sol. (a)* **204**, 3078 (2007).
15. W. Chen, D. Qi, X. Gao and A. T. S. Wee, *Progress in Surface Science* **84**, 279 (2009).
16. C. Liu and A. J. Bard, *Nature Mater.* **7**, 505 (2008).
17. Yu. Ya. Gurevich and Yu. V. Pleskov, *Russ. J. Electrochem.,* (Transl. of *Elektrokhimiya*) **18**, 1477 (1982).
18. S. Trasatti, *Pure and Appl. Chem.* **58**, 955 (1986).
19. H. Reiss and A. Heller, *J. Phys. Chem* **89**, 4207 (1985).
20. H. Reiss, *J. Phys. Chem.* **89**, 3783 (1985).
21. A. W. Adamson, *Physical Chemistry of Surfaces,* John Wiley, NY, 4th edition, 1982.
22. J. J. Mareš, P. Hubík, J. Krištofík, J. Ristein, P. Strobel, L. Ley, *Diamond and Rel. Mat.* **17**, 1356 (2008).
23. J. Ristein, M. Reidel, M. Stammler, B. F. Mantel and L. Ley, *Diamond and Related Mat.* **11**, 359 (2002).
24. G. Sumanasekeran and S. Desai, personal communication, 2009.
25. V. Chakrapani, J. C. Angus, A. B. Anderson, G. Sumanasekera, "Diamond Electronics Symposium," *Mat. Res. Soc. Symp. Proc.* **2007**, *956,* paper J15-01.

Device Applications

(3) Radiation Detectors

Mater. Res. Soc. Symp. Proc. Vol. 1203 © 2010 Materials Research Society 1203-J17-35

Dosimetric Assessment of Mono-Crystalline CVD Diamonds Exposed to Beta and Ultraviolet Radiation

M. Pedroza-Montero, R. Meléndrez, S. Preciado-Flores, V. Chernov, M. Barboza-Flores.
Centro de Investigación en Física, Universidad de Sonora, Apartado Postal 5-088, Hermosillo, Sonora, 83190 México.

ABSTRACT

Polycrystalline and mono-crystalline CVD diamonds have been investigated in relation to radiation dosimetry applications. In this work we report results on the thermoluminescence (TL), afterglow (AG) and dosimetric performance on two mono-crystalline CVD diamonds containing boron and silicon as doping materials. The samples were exposed to beta (Sr^{90}/Y^{90}) in the dose range of 0.07-8.26 Gy and UV light in the range of 200-400 nm, followed in both cases, by TL and AG read-outs. The boron doped sample exhibited one main TL peak at 130 °C and some overlapped peaks around 250 °C and the silicon doped samples exhibited two TL peaks around 148 and 286 °C after the crystals were subjected to beta radiation. UV radiation exposed samples showed two main TL peaks around 139 °C for boron doped, with overlapped components in the high temperature side, and at 220 and 355 °C for silicon doped samples. The integrated TL and AG intensities reached saturation around to 3.0 and 1.0 Gy in boron and silicon doped samples, respectively. The AG signal from boron doped samples reached saturation for around 60 s of 230 nm UV light irradiation and the silicon doped sample showed a linear response up to 10 minutes of 300 nm UV exposure with no apparent saturation for higher irradiation times. The TL/AG behavior of the present CVD diamond indicates the promising applications of these materials as TL/AG dosimeter for ionizing and UV radiation.

INTRODUCTION

A diversity of polycrystalline, homoepitaxial, heteroepitaxial CVD diamonds have been exhaustively studied as radiation and TL detectors for many research groups, due to their excellent features such as high radiation hardness, high temperature operation capability, chemical inertness and tissue equivalence that make them suitable for biomedical applications [1-6]. However the TL and radiation detector performance of these synthetic diamonds strongly depends on the film quality of the CVD diamond which also depend on the doping component, precursor gas and reactor conditions used.

Recently, high quality mono-crystalline and homoepitaxial diamonds were synthesized by using different growth methods showing superior properties in terms of homogeneity of response, carrier mobility and trapping times [5,7, 8]. These improvements caused a remarkable progress in the study of several applications, including electronic and dosimetry devices development [9,10]. However, few results have been reported in relation to mono-crystalline CVD diamonds as thermoluminescence radiation detectors and dosimeters.

In the present study, we report on the thermoluminescence and afterglow responses of two mono-crystalline CVD diamonds doped with boron and silicon exposed to beta and UV radiation.

EXPERIMENT

Two homoepitaxial diamonds were grown by Microwave Plasma Chemical Vapor Deposition method (MWPCVD) on off-axis Ib HPHT diamond substrates and a microwave power of 1400 W. The free standing samples, obtained by removing the substrate, were synthesized using boron (1.04 mm average thickness) and silicon (0.8 mm average thickness) components. The gas pressure in the chamber was of 30 mbar with a methane concentration of 5% and 11% in H_2, respectively. Both diamonds samples were subjected to an annealing treatment up to 800 °C during 30 min previously of being exposed to ionizing and non ionizing radiation.

The radiation with beta particles, TL and the afterglow (AG) measurements were obtained using a Risø TL/OSL system model TL/OSL DA-15, equipped with a 33 mCi ^{90}Sr beta source and a dose rate of 4.13 Gy/min, in the range of 0.07-8.26 Gy. UV exposures were performed in the range of 200-400 nm, using a xenon lamp and a monochromator Oriel model 77250. The arrangement was coupled through an optic fiber to the Risø equipment.

The TL readouts were obtained up to 350 °C with a heating rate of 5 °C /s. The AG curves were recorded immediately after the radiation exposure in darkness conditions and at room temperature during 60 s. Both TL/AG dosimetric techniques were performed for several times to measure reproducibility.

DISCUSSION

Both CVD samples were subjected to beta radiation in the range of 0.07-8.26 Gy, immediately after each irradiation the afterglow was measured during 60 seconds in total darkness and at room temperature. Figure 1 shows a comparison of the afterglow curves decay for the CVD diamond doped with boron and silicon, at different doses. The monotonous decay is observed in all cases and the boron doped sample showed a higher AG intensity.

On the other hand, figure 2 presents the thermoluminescence glow curves at the same doses depicted in figure 1. The TL glow curve displays two pronounced maxima at 130 °C (with an overlapped TL structure around 325 °C), and at 142 and 280 °C for the boron and silicon doped samples, respectively. The observed differences in the TL glow pattern may be attributed to the doping content. It is well know that there is an excess of hydrogen in monocrystalline CVD diamonds which comes from the chemical growth environment process. The defects associated to boron and silicon doping and hydrogen may take an important participation in the TL signal [8,11]. However, we can not let aside the possibility of the presence of other elements as tracing impurities, which may generate some TL peak maxima pattern changes similar to those observed in figure 2 and reported elsewhere in monocrystalline CVD diamond films exposed to beta particles [4].

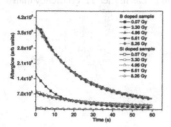

Figure 1. Afterglow decay curves of boron and silicon doped CVD diamond after β irradiation at different doses.

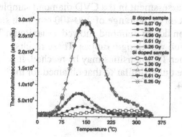

Figure 2. Thermoluminescence glow curves of boron and silicon doped CVD diamond after β irradiation and different doses.

The boron and silicon doped diamond samples were subjected to ultraviolet radiation of 230 and 300 nm, respectively. The normalized TL glow curves at different exposure times are show in the figure 3. The silicon doped sample was multiplied by a factor of 20 to make possible a comparison in the same figure scale. The boron doped sample depicts TL glow peak with maxima at about 140 °C and an overlapped TL glow structure around 330 °C. In the case of the silicon doped sample the TL glow are not well defined due to its low TL efficiency as compared to boron doped diamond, although some TL structure may be observed around 220 and 345 °C.

There are considerable differences in the TL glow curves in samples subjected to ionizing and UV radiation. The TL efficiency in boron doped diamond is higher than silicon doped diamond for UV irradiation. However, silicon doped diamond has higher TL efficiency than doped samples for β irradiation.

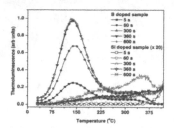

Figure 3. Normalized TL glow curves for boron and silicon doped CVD diamond UV irradiated at the indicated time exposures.

To perform the ultraviolet radiation dosimetry assessment in the CVD diamond samples, it was necessary to obtain the AG and TL creation spectra in the range of 200-400 nm, with steps of 10 nm as displayed in figure 4. The boron and silicon doped diamond showed a maximum TL efficiency in the 200-250 nm and 270-310 nm, excitation wavelength ranges. These results provide us the excitation wavelength at which the higher TL intensities may be reached. It is important to mention that the AG creation spectra were quite similar to that obtained for the thermoluminescence creation spectra shown in figure 4.

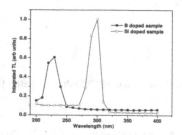

Figure 4. Thermoluminescence creation spectra for boron and silicon CVD diamond.

The TL dose response in shown in figure 5 using a normalized scale in which the boron doped sample response is multiplied by a factor of 3. It is clearly observed that the TL signal saturates at 1.0 Gy and 3.0 Gy for silicon and boron doped diamond, respectively.

208

about it really could show TL peaks in the 30 – 300 °C temperature range, in the TL linearity region is great.

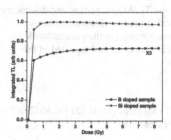

Figure 5. Normalized TL dose response for boron and silicon doped CVD diamond exposed to β radiation.

Figure 6 present the AG dose response after UV irradiation for different time exposure in 5-600 seconds range. The figure also displays the integrated AG signal reproducibility for three cycles. It is evident the existence of some linear dose response for different irradiation times with a very good reproducibility properties in both types of doped CVD diamond samples. The boron doped specimen has a higher AG efficiency and no apparent saturation for 600 s of irradiation.

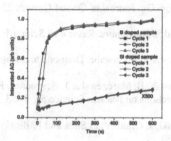

Figure 6. Integrated AG intensity as a function of UV irradiation time exposures. The picture shows signal reproducibility in three consecutive cycles.

CONCLUSIONS

The TL/AG properties in boron and silicon doped CVD diamond samples indicate the possibility of using high quality mono-crystalline and homoepitaxial CVD diamonds as TL radiation detectors and dosimeters. The CVD diamond samples may be suitable for application involving ionizing and UV radiation applications in the medical and environmental fields.

Although, both samples show TL peaks in the low temperature side around145 °C the dosimetric behavior is quite good.

The boron and silicon dopants are clearly affecting the TL/AG properties and dosimetric behavior and a deeper study on the nature of the trapping and recombination mechanisms responsible for the observed TL/AG is necessary. The results obtained in the present work are comparable to those already obtained in polycrystalline CVD diamonds and reported in the literature.

ACKNOWLEDGEMENTS

We are grateful for the support from by CONACYT (México) grants 53149, 36521, 37641, 32069 and PROMEP and PIFI programs from SEP (México).

REFERENCES

1. A. Balducci, Y. Garino, A. Lo Giudice, C. Manfredotti, Marco Marinelli, G. Pucella and G. Verona-Rinati, Diamond & Related Materials **15**, 797 (2006).
2. M. Barboza-Flores, M. Schreck, S. Preciado-Flores, R. Meléndrez, M. Pedroza-Montero and V. Chernov, Phys. Stat. Sol. (a) **204**, 3047 (2007).
3. C. Descamps, D. Tromson, M. J. Guerrero, C. Mer, E. Rzepka, M. Nesladek and P. Bergonzo, Diamond & Related Materials **15**, 833 (2006)
4. S. Preciado-Flores, M. Schreck, R. Meléndrez, V. Chernov, M. Pedroza-Montero and M. Barboza-Flores, Phys. Stat. Sol. (a) **203**, 3173 (2006).
5. C. Manfredotti, A. Lo Giudice, S. Medunic, M. Jaksic and E. Colombo, Mater. Res. Soc. Symp. Proc. **908E**, 20.1 (2006).
6. Minglong Zhang, Yibean Xia, Linjun Wang, Beibei Gu, Journal of Crystal Growth **227**, 382 (2005)
7. J. H. Kaneko, T. Teraji, Y. Hirai, M. Shiraishi and S. Kawamura, Review of Scientific Instruments **75**, 3581 (2004).
8. K. Iakoubovskii, A. Stesmans, K. Suzuki, J. Kuwabara, A. Sawabe, Diamond and Related Materials **12**, 511 (2003).
9. A. Galbiati, S. Lynn, K. Oliver, F. Schirru, T. Nowak, B. Marczewska, J. A. Dueñas, R. Berjillos, I. Martel and L. Lavergne, IEEE Transactions on Nuclear Science **56**, 1863 (2009).
10. J. Morse, M. Salomé, E. Berdermann, M. Pomorski, J. Grant, V. O'Shea and P. Ilinski, Mater. Res. Soc. Symp. Proc. **1039**, P06-02 (2008).
11. J. Chevallier, D. Ballutaud, B. Theys, F. Jomard, A. Deneuville, E. Gheeraert and F. Pruvost, Phys. Stat. Sol. (a) **174**, 73 (1999).

Mater. Res. Soc. Symp. Proc. Vol. 1203 © 2010 Materials Research Society   1203-J17-43

Simulations of Charge Gain and Collection Efficiency from Diamond Amplifiers

Dimitre A. Dimitrov[1], Richard Busby[1], John R. Cary[1], Ilan Ben-Zvi[2,3], John Smedley[2], Xiangyun Chang[2], Triveni Rao[2], Jeffrey Keister[2], Erik Muller[2,3], and Andrew Burrill[2]
[1]Tech-X Corporation, Boulder, CO, U.S.A.
[2]Brookhaven National Laboratory, Upton, NY, U.S.A.
[3]Stony Brook University, Stony Brook, NY, U.S.A.

ABSTRACT

A promising new concept of a diamond amplified photocathode for generation of high-current, high-brightness, and low thermal emittance electron beams was recently proposed and is currently under active development. To better understand the different effects involved, we have been developing models, within the VORPAL computational framework, to simulate secondary electron generation and charge transport in diamond. The implemented models include inelastic scattering of electrons and holes for generation of electron-hole pairs, elastic, phonon, and charge impurity scattering. We will discuss these models and present results from 3D VORPAL simulations on charge gain and collection efficiency as a function of primary electron energy and applied electric field. The implemented modeling capabilities already allow us to investigate specific effects and compare simulation results with experimental data.

INTRODUCTION

To address the need for high average-current, high brightness electron beams in current and future accelerator-based systems (e.g. electron cooling of hadron accelerators, energy-recovery linac light sources, and ultra-high power free electron lasers), a new design for a photoinjector with a diamond amplifier was proposed [1] and is under intensive development [2,3]. The technologically important material properties of diamond have been actively researched for emission [4] and detector applications. Moreover, diamond is currently being investigated to develop high-flux x-ray monitors for synchrotron beam lines [5-7].

The new photoinjector concept has important advantages [1] compared to existing metallic and semi-conductor photocathodes. The idea of its operation is to first generate a primary electron beam using a conventional photocathode and inject it into diamond. The primary electrons scatter in diamond, generating a cascade of secondary electrons. More recently, photons are also being considered in experiments as a source for generation of secondary electrons instead of using primary electrons.

The secondary electrons drift through the diamond under the acceleration of an applied electric field. The transported secondary electrons are emitted from a diamond surface with a negative electron affinity into the accelerating cavity of an electron gun. Over two orders of magnitude charge amplification (number of generated secondary electrons emitted relative to the number of primary electrons used as input) could potentially be achieved using this approach.

In order to better understand the phenomena involved in electron amplification and emission from diamond, we have been implementing models [8,9], within the VORPAL [10] computational framework, for simulation of secondary electron generation and charge transport. Here, we report results on gain obtained by considering only the loss of electrons due to electron cloud expansion near the diamond surface where primary electrons enter.

We compare simulation results with results from experiments [2,3] on electron gain when using primary electrons to generate the secondaries. The implemented modeling capabilities allow us to consider, at least qualitatively, physical properties (collection efficiency, in particular) related to how diamond responds to x-ray photons. We will also discuss initial results from our simulations on studying collection efficiency (CE).

THEORY

When a primary electron with sufficiently high energy enters into diamond, it undergoes inelastic scattering leading creation of electron-hole (e-h) pairs. Moreover, some of the secondary electrons and the holes are created with sufficiently large energies to produce additional e-h pairs.

These inelastic processes continue until the free charge carriers (electrons in the conduction band and holes in valance band) no longer have enough energy to cause the transition of electrons from the valance to the conduction band. The minimum energy for such transitions is equal to the energy gap in diamond, $E_G = 5.47$ eV, at room temperature. However, the mean free time between such inelastic scattering events (it can be obtained from the total cross section for this process [11,8]) increases towards infinity when the energy of the charge carrier decreases towards E_G. Effectively, when the energy of the free carriers is close to 10 eV, the further relaxation of the electron energies towards the bottom of the conduction band is due to low energy inelastic scattering with phonons (and similarly for holes).

For the results presented here, we used our implementation of Ashley's model [11] for the inelastic scattering leading to e-h generation at 300 K. For the low-energy inelastic scattering with optical and acoustic phonons, we implemented Monte Carlo algorithms based on the models given by Jacoboni and Reggiani [12]. Our electron-phonon scattering implementation leads to electron drift velocities that are in agreement with band structure calculations and with experimental data [13]. We have coded a general Monte Carlo algorithm to handle all scattering processes. In between scattering events, the particle-in-cell algorithms in VORPAL [10] provide the capabilities to self-consistently move particles interacting with electromagnetic fields.

From the time a primary electron starts moving in diamond, the secondary electron generation (initiated by the primary electron) is close to complete after 200 fs. The code switches electrons and holes to phonon scattering only, once their energy falls below 11 eV during the first 400 fs. The simulations for studying electron gain starts with a primary entering the diamond surface set at the x=0 side of the simulation domain.

The simulations for evaluating CE start with a distribution of electrons (inside diamond) located at different depths from a surface considered as sink for particles. The initial part of the simulations is done with a time step Δt small enough (of the order of 0.01 fs) to resolve the scattering rate of the high-energy charge carriers.

At 400 fs, we dump the simulation state and restart with all particles switched to inelastic scattering with phonons only. These scattering processes have several orders of magnitude larger mean free time. This allows selecting a larger time step after the restart. We have used up to $\Delta t = 2$ fs. The cell size was varied in the range from 0.1 to 2 μm. In all simulations, the cell sizes were equal along the x, y and z directions.

DISCUSSION

Charge Gain

In the simulations, we estimate gain by counting the number of free electrons that successfully separate from the domain of the diamond surface where the secondary electrons are generated. These results can be compared, at least qualitatively, with the gain measured from transmission mode [2] experiments (metal contacts on opposite diamond surfaces are used to create an electric field of up to several MV/m). Primary electrons are injected through one of the metal contact surfaces (at $x = 0$ in the simulations with the applied field in the x direction). The generated secondary electrons and holes move in opposite directions due to the applied field but also expand diffusively due to scattering. In the initial stage of the charged cloud evolution, some of the electrons will reach into the metal contact at the $x = 0$ surface and will be lost from diamond. The free electrons that are successfully separated from the domain of generation drift towards the opposite metal contact. We count these electrons as the realized gain. The hole cloud also expands diffusively but it drifts towards the initial metal contact surface and is extracted there. We model this behavior by using a sink boundary condition at the $x = 0$ side of the simulation box. The primary electrons are started at this side with an initial velocity along the positive x axis.

In Fig. 1, we show the time evolution of the number of electrons and holes for 3 and 0.1 MV/m applied electric fields in diamond. As expected, the rate at which electrons drift away from the diamond surface increases when increasing the field. The less amount of time the electrons are close to the metal contact surface, the less number of electrons are lost from diamond, thus, leading to higher gain.

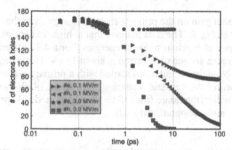

Figure 1: The evolution of the electron and hole numbers vs time is as expected – these numbers reach saturation faster when increasing the applied electric field.

We compare gain results from the simulations to experimental data [2] for primary electron energies of 0.7, 1.7, 2.7, 3.7, and 4.7 keV on the left plot in Fig. 2. The simulations show the same overall dependence on the applied field as the experiments. The results from the simulations and the experiments are in good agreement for the 2.7 keV case. They agree qualitatively at the lower and higher energies investigated. For the 3.7 and 4.7 keV cases, the simulations predict 16 and 25 % higher gain than observed in the experiments.

The higher gain seen in the simulations for these two cases could be due to two reasons. In the simulations, the gain was determined by the number of electrons that are successfully separated from the metal contact at the diamond surface near which the generation happens. Any additional electron loss during the propagation to the other particlesmetal contact is not taken into account (e.g., due to trapping or recombination with holes). The gain in the experiments is determined by the total transmitted charge collected at the opposite metal contact. Secondly, the gain in the experiments was determined [2] assuming that the primary electrons lose a constant amount of energy when going through the metal contacts before entering in diamond. However, this loss is generally dependent on the energy of the particles.

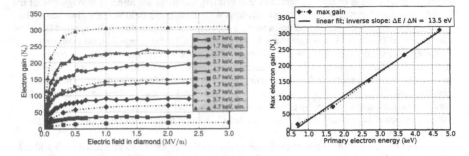

Figure 2: Comparison of electron gain from transmission experiments [2] with simulation results (left) show qualitative agreement overall and quantitative for the middle of the energy range. The maximum number of electrons separated from the generation domain (right) shows two order of magnitude charge gain for primary electron energy larger than 2 keV.

The dependence of the maximum gain on the primary electron energy from the simulations is shown in the right plot of Fig. 2. The data indicate that at higher energies, the diffusive motion does not affect the gain as much as at lower energies. For a 4.7 keV primary electrons, the average energy to generate a secondary electron is about 15 eV. However, for a 700 eV primary electron, it is 38 eV. Note that simulations started with a primary electron in bulk diamond (using the Ashley model for the inelastic scattering) give about 14 eV average e-h pair generation energy. This value is within the range from 10 to 17 eV reported [11] previously by different experimental, theoretical, and computational studies.

<u>**Collection Efficiency**</u>
The currently implemented modeling capabilities also allow us to directly estimate the CE as a function of the depth at which photons are absorbed since we can start the simulations with an initial distribution of e-h pairs at any given depth inside diamond. The collection efficiency of a detector is defined as the ratio $\eta = Q_c/Q$ of collected charge Q_c to the total charge Q generated in response to the absorbed photons. The obtained CE vs depth (measured from the surface photons enter into diamond) for 600 eV photons and several values of the applied electric field are shown in the left plot of Fig. 3. In the simulations here, we determine the CE by measuring the number of electrons successfully separate from the domain of generation (after the loss of electrons to the metal contact has completed) and separately the total number of electrons generated by the input energy of the absorbed photons. As expected, the deeper the absorption occurs in diamond and the higher the applied field, higher CE is obtained.

214

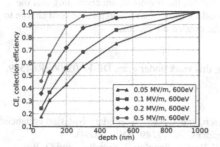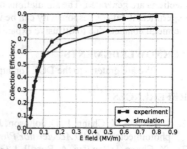

Figure 3: Simulations allow direct investigation of CE vs absorption depth – results for different depths and fields for 600 eV photons (left plot). Experimental data on CE vs field for 600 eV photons are in overall qualitative agreement with the simulation results (right plot).

In Fig. 3 (right plot), we compare simulation and experimental data on the dependence of the CE vs electric field for 600 eV photons. Overall, they are in qualitative agreement (and in quantitative at lower fields). However, the most important insight from these simulations is new understanding about the experimental data on responsivity vs photon energy. The results from the VORPAL simulations show that the observed responsivity (S), measured in A/W, can be described by the expression

$$S(v) = \frac{1}{W_C} \exp\left(-\frac{t_M}{\lambda_M(v)}\right)\left(1 - \exp\left(-\frac{t_C}{\lambda_C(v)}\right)\right) CE(v, E), \qquad (1)$$

where W_C is the mean energy to create an e-h pair, t_M and t_C are respectively the thickness lengths of the metal contact layer and the diamond sample; λ_M and λ_C are the corresponding photon energy-dependent absorption lengths. The energy and field dependent collection efficiency function, $CE(v, E)$, is obtained by an empirical fit to the VORPAL simulations data. The VORPAL results show that the loss of charge carriers to the metal contact (due to the diffusive expansion of the charge cloud) provide an explanation for the observed data. This charge loss effect is most important for low photon energies, less than 1 keV. For these energies, the e-h pairs are generated close to the metal contact and the diffusive expansion allows some charge carriers to reach the metal contact. The details of this approach to describe the observed data are given by Keister *et al.* [14].

CONCLUSIONS

We reported here initial simulation results on charge gain and CE from diamond. Higher gain can be achieved by increasing the primary electron energy and the magnitude of the applied electric field. However, increasing the field will lead to increasing the effective temperature of the drifting electrons [12,13,9], thus, increasing the thermal emittance of electron beams. At the intermediate primary electron energies considered, the gain from the simulations agrees well with experimental data. Overall, the agreement is qualitative. However, in the simulations we have considered only loss due to the diffusive motion near the diamond surface where secondary electrons are generated. The simulations on CE provided the important understanding that the experimental data on responsivity can be understood by taking into account the loss of charge to the metal contacts due to diffusive expansion of the created e-h clouds.

ACKNOWLEDGMENTS

We are grateful to the DOE for supporting this work under grant DE-FG02-06ER84509.

REFERENCES

1. I. Ben-Zvi, X. Chang, P. D. Johnson, J. Kewisch, and T. Rao. "Secondary emission enhanced photoinjector". *C-AD Accelerator Physics Report C-A/AP/149*, BNL, (2004).
2. X. Chang, I. Ben-Zvi, A. Burrill, J. Grimes, T. Rao, Z. Segalov, J. Smedley, and Q. Wu, "Recent Progress on the Diamond Amplified Photo-cathode Experiment". In *Particle Accelerator Conference (PAC07)*, pp. 2044–6, IEEE (2007).
3. X. Chang, I. Ben-Zvi, A. Burrill, J. Kewisch, E. Muller, T. Rao, Z. Segalov, J. Smedley, E. Wang, Y. Wang, and Q. Wu, "First Observation of an Electron Beam Emitted From a Diamond Amplified Cathode", In *PAC09*, IEEE (2009).
4. J. E. Yater and A. Shih, "Secondary electron emission characteristics of single-crystal and polycrystalline diamond", *J. Appl. Phys.* **87**, pp. 8103-12 (2000) and the references by J. Yater *et al.* there in.
5. J. Keister and J. Smedley, "Single crystal diamond photodiode for soft x-ray radiometry", *Nucl. Instr. Meth. Phys. Res. A*, **606**, pp. 774–9 (2009).
6. J. Smedley, J. W. Keister, E. Muller, J. Jordan-Sweet, J. Bohon, J. Distel, B. Dong, "Diamond Photocathodes for X-ray Applications", these proceedings.
7. J. Bohon, J. Smedley, E. Muller, and J. W. Keister, "Development of Diamond-Based X-ray Detection for high Flux Beamline Diagnostics", these proceedings.
8. D. A. Dimitrov, R. Busby, D. L. Bruhwiler, J. R. Cary, I. Ben-Zvi, T. Rao, X. Chang, J. Smedley, and Q. Wu, "3D Simulations of Secondary Electron Generation and Transport in a Diamond Amplifier for Photocathodes", In *PAC07*, pp. 3555–7, IEEE (2007).
9. R. Busby, D. A. Dimitrov, J. R. Cary, I. Ben-Zvi, X. Chang, J. Keister, E. Miller, T. Rao, J. Smedley, and Q. Wu, "3D Simulations of Secondary Electron Generation and Transport in Diamond Electron Beam Amplifiers", In *PAC09*, IEEE (2009).
10. C. Nieter and J. Cary, "VORPAL: a versatile plasma simulation code", *J. Comput. Phys.*, **196**, pp. 448-73 (2004).
11. B. Ziaja, R. A. London, and J. Hajdu, "Unified model of secondary electron cascades in diamond", *J. Appl. Phys.*, **97**, pp. 064905–1/9 (2005).
12. C. Jacoboni and L. Reggiani, "The Monte Carlo method for the solution of charge transport in semiconductors with applications to covalent materials", *Rev. Mod. Phys.*, **55**, pp. 645–705 (1983).
13. T. Watanabe, T. Teraji, T. Ito, Y. Kamakura, and K. Taniguchi, "Monte Carlo simulations of electron transport properties of diamond in high electric fields using full band structure", *J. Appl. Phys.*, **95**, pp. 4866–74 (2004).
14. J. W. Keister, J. Smedley, D. A. Dimitrov, and R. Busby, "Charge Collection and Propagation in Diamond X-Ray Detectors", submitted.

Mater. Res. Soc. Symp. Proc. Vol. 1203 © 2010 Materials Research Society   1203-J17-19

Electronic Impact of Inclusions in Diamond.

Erik M. Muller[1], John Smedley[2], Balaji Raghothamachar[1], Mengjia Gaowei[1], Jeffrey W. Keister[3], Ilan Ben-Zvi[4], Michael Dudley[1], and Qiong Wu[4]

[1]Department of Materials Science and Engineering, Stony Brook Universtity, Stony Brook, NY 11794, U.S.A.
[2]Instrumentation Division, Brookhaven National Laboratory, Upton, NY 11973, U.S.A.
[3]National Synchrotron Light Source II, Brookhaven National Laboratory, Upton, NY 11973, U.S.A.
[4]Collider-Accelerator Division, Brookhaven National Laboratory, Upton, NY 11973, U.S.A.

ABSTRACT

X-ray topography data are compared with photodiode responsivity maps to identify potential candidates for electron trapping in high purity, single crystal diamond. X-ray topography data reveal the defects that exist in the diamond material, which are dominated by non-electrically active linear dislocations. However, many diamonds also contain defects configurations (groups of threading dislocations originating from a secondary phase region or inclusion) in the bulk of the wafer which map well to regions of photoconductive gain, indicating that these inclusions are a source of electron trapping which affect the performance of diamond X-ray detectors. It was determined that photoconductive gain is only possible with the combination of an injecting contact and charge trapping in the near surface region. Typical photoconductive gain regions are 0.2 mm across; away from these near-surface inclusions the device yields the expected diode responsivity.

INTRODUCTION

Recent advances in the quality of synthetically grown single crystal diamond have enabled much research into diamond based electronics. The high mobility, high thermal conductivity, large bandgap (making it solar blind) and radiation hardness of diamond combine a unique set of properties making devices such as high-flux radiation detectors feasible [1]. The purity and crystalline quality of chemical-vapor deposition (CVD) grown diamonds has improved to the extent that charge trapping effects do not forbid their use in electronic applications. However, charge trapping has not been eliminated from devices fabricated using diamond, and, as will be shown, electron trapping near the surface of the hole injecting electrode results in photoconductive gain. Measurements on polycrystalline diamond X-ray detectors have shown regions of photoconductive gain which may be related to crystalline structure or the presence of impurities [2]. By performing similar measurements utilizing single crystal CVD diamond in conjunction with X-ray topography we have been able to identify crystalline defects which result in regions of photoconductive gain.

EXPERIMENT

Two measurement techniques are combined to identify the source of photoconductive gain regions in diamond X-ray photodiodes. First, X-ray topography is used to identify potentially electrically active defects in the crystal. Second, 2D responsivity maps of planar X-ray photodiodes are collected. The resulting two images are then compared to reveal any correlations between electrical features and crystalline defects. The diamond samples used in these experiments are electronic grade (< 5 ppb nitrogen impurity content), single crystal (001) purchased from Element Six. The nominal size is 4 mm × 4 mm × 0.3 mm with a <110> edge normals.

White beam X-ray topography

White beam X-ray topography images were obtained at beamline X19C at the National Synchrotron Light Source (NSLS). With a 2.5 GeV 100 mA electron beam in the synchrotron accelerator, the flux at the beamline is 4.16×10^{12} ph/sec for 4.9 keV photons. A 4-25 keV nearly parallel white X-ray beam is incident on the sample in transmission geometry. The thin diamond sample is oriented in the incoming synchrotron white beam on a precision goniometer so as to record the various $\overline{2}20$ and 111 reflections on a 20 cm × 25 cm film placed in the transmission position perpendicular to the incident beam at 10 cm from the crystal. The direct beam is blocked after the sample to only allow the diffraction images appear on the film. See figures 1 and 2 for typical topographs from these single crystal CVD diamonds (note that the two topographs come from different diamonds). Because of the high X-ray intensity, the exposure time must be kept within several seconds in order to achieve clear images on the film. X-ray topography has previously been done on CVD diamond (optical grade) grown on type Ib diamond showing mainly linear dislocations originating from the growth substrate [3]. While the electronic grade single crystal diamonds exhibit many defects, they have a qualitatively higher crystal quality than the optical grade (<5 ppm nitrogen impurity) diamonds. X-ray topographs of optical grade diamonds are completely black when using the same short exposure times used for the electronic grade diamonds indicating a much larger defect density.

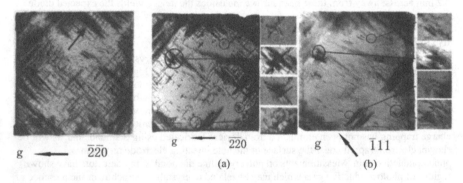

g ——— $\overline{2}\overline{2}0$ g ——— $\overline{2}\overline{2}0$ g $\diagdown$ $\overline{1}11$
                          (a)   (b)

Figure 1. X-ray topograph ($g = \overline{2}\overline{2}0$) showing dislocation slip bands.

Figure 2. X-ray topographs indicating the defect features associated with inclusions (see insets for magnified view); (a) g = $\overline{2}\overline{2}0$; (b) g = $\overline{1}11$.

2D responsivity mapping

Two dimensional responsivity maps are obtained by raster scanning an X-ray microbeam over a planar diamond X-ray photodiode and plotting the current at each position. These devices are fabricated by the following steps. First, the diamonds are cleaned in an acid bath to remove any metal or non-diamond contamination. The diamonds are then placed under a UV lamp in air for many hours to further clean the diamond and leave the surface with an oxygen termination. Platinum contacts, 30 nm thick, are sputter coated over a 3 mm diameter area on both sides of the diamond. The sample mount consists of two copper jaws which make electrical contact with the two platinum contacts.

The responsivity maps shown are acquired at beamlines X15 and X6B at the NSLS. The X6B beamline provides a focused X-ray beam which can be closed down to ~10 μm using slits and still provide sufficient signal in the diamond X-ray photodiode. All maps shown are done with photon energy of 19keV. At this energy, the X-ray attenuation length in diamond is ~7 mm, which is much thicker than the diamond. The result is that a column of charge carriers are generated throughout the diamond thickness, providing a constant source of active detrapping. This avoids the need to use a pulsed biasing scheme for detrapping that is needed when using soft X-ray to generate charge carriers [4].

A bias is placed on the X-ray incident side and the current is measured on the opposite side with a Keithley Electrometer, model number 6514. Typically a bias of ±100V is used and is well above diode saturation where full charge collection is achieved. By using an x-y translation stage, the sample is raster scanned in the X-ray microbeam. A silicon photodiode is placed several inches downstream of the diamond sample. Contrast can be seen in the silicon photodiode signal due to X-ray photon loss in the diamond. Though small, the photon loss provides contrast outlining the diamond position. This is important to allow for proper alignment of the 2D responsivity maps with the X-ray topographs.

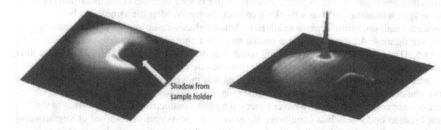

Figure 3. Responsivity map showing uniform diode response over the entire metalized region.

Figure 4. Diamond exhibiting a small region of photoconductive gain. Expected diode response is seen everywhere else on the metalized region.

Responsivity maps are acquired with the diamond X-ray photodiodes as prepared with the oxygen terminated surface and also after annealing the device above 600°C. Figure 3 shows a typical responsivity map on a diamond X-ray photodiode before annealing and has a uniform predictable diode response over the entire metalized area. This response map was acquired with a beam spot of 0.250 mm and the entire acquisition time of the image is approximately 20 minutes. More detailed images have been acquired for other diamonds prior to annealing and show the same uniform diode response. Figure 4 show the response map of the same diamond after annealing the sample above 600°C. In this case the X-ray spot size was 50 μm and the entire image acquisition time was approximately 3 hours. All current measurements are taken with the electrometer in the "normal" or "high accuracy" speed mode, corresponding to a 17 ms and a 167 ms integration time respectively.

DISCUSSION

The recorded topographs shown in Fig. 1 and Fig. 2 have several broad dark bands running diagonally on the image mostly emanating from the crystal edges and running towards the crystal interior. These correspond to slip bands comprising dislocation pile-ups. Additionally, topographs in Fig. 2 show small configurations of line dislocations which appear to emanate either from the surface (original substrate-crystal interface) or from a defect region lying within the bulk of the sample. On the $\bar{1}11$ reflection (Fig. 2(b)), these groups of line dislocations appear to lie almost along the <001> growth direction and deviate along the <110> directions as previously observed in CVD diamond [8]. Such growth dislocations are commonly generated by secondary phase or inclusions (impurities, precipitates, voids, etc.) either in the substrate or formed inside the crystal during the growth process [7]. The net Burgers vector of such groups of growth dislocations so nucleated must be zero (see for example [7] and [9]). On the topographs, these features have different lengths (see Fig. 2(b) insets). The longest features likely originated at the substrate-crystal interface and run the entire thickness while the shorter features originate within the crystal at, for example, an inclusion. These topographs are overlaid with the 2D responsivity maps to identify any features which correlate with features in the responsivity.

In the case where the sample has not yet been annealed, the responsivity is flat and featureless, see figure 3. The expected diode response is seen over the entire metalized region. The oxygen termination creates a blocking contact. However, after the sample has been annealed, small prominent regions exhibiting photoconductive gain appear in the responsivity map, see figure 4. A cross-section line profile through a photoconductive region is shown in figure 5. These features have a typical full width half maximum of is ~200 μm. This size is likely an overestimation of the true extent of the photoconductive region as there is a noticeable persistent current flowing even after the X-ray beam is moved from the region. The presence of these photoconductive regions depends on the polarity of the bias applied (whether holes or electrons are collected). We have never observed a photoconductive region appearing in the same location under both bias conditions. By annealing, the oxygen is desorbed or migrates into the platinum electrode and the Schottky barrier at the contact is either greatly reduced or removed completely, allowing charge injection [10].

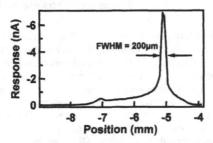

Figure 5. Typical profile across a photocondutive region.

The responsivity maps are overlaid with the topographs demonstrating that every region where there is photoconductive gain, there are defect features associated with an inclusion, see figure 6. However, not all such features appear to be cause photoconductive gain. This correlation between such defect features and photoconductive gain has been confirmed with several diamonds. We have also physically flipped the diamond so that the incident side becomes the exiting side; this effectively changes the surface at which holes or electrons are collected. In this case the photoconductive regions first observed under positive bias would now be observed in negative bias and vice versa. This has lead to the conclusion that the position in depth of the inclusion determines whether or not a particular inclusion results in photoconductive gain. More specifically, inclusions/voids near the hole-injecting surface will result in photoconductive gain. This explanation does not exclude the possibility that electrons are trapped elsewhere in the diamond (which is likely the case), however, the electric field generated by these trapped electrons does not cause hole injection. Possibilities include small amounts of trapping at dislocations or impurities as well as inclusion buried deeper toward the center of the diamond.

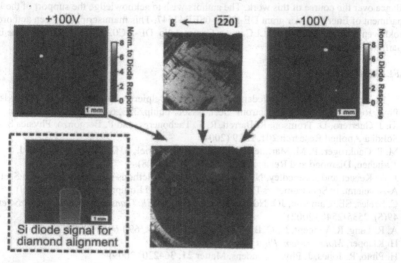

Figure 6. Topographs and responsivity maps overlaid to show correlation of photoconductive regions with inclusions associated features. Silicon diode signal shown is used to align the responsivity maps with the topographs.

CONCLUSIONS

Regions of photoconductive gain are observed to correlate with inclusions/voids associated defect features in single crystal electronic grade diamond. Photoconductive gain is only observed for near-surface inclusions in diamond X-ray photodiodes created with injecting contacts. The adverse affect of near surface inclusions can be avoided with the creation of robust blocking contacts or by avoidance of annealing the diamond, i.e. for use in low flux operation where the diamond will not be heated. The nature of the inclusions is not known and they are identified only by the three-dimensional strain fields that appear in the topographs. There is a range of possibilities for their origin including voids, foreign material to non-diamond carbon. Efforts to definitely identify the depth of the inclusions which support photoconductive gain are ongoing and include X-ray topography at grazing incidence and section topography. Further work needs to be done to obtain detailed crystallographic identification of the inclusions.

ACKNOWLEDGMENTS

X-ray topography experiments were carried out at the Stony Brook Synchrotron Topography facility (Beamline X-19C) at the NSLS, Brookhaven National Laboratory. The authors wish to thank John Walsh for design and fabrication of sample mounts and Xiangyun Chang for assistance with metallization and Bin Dong for technical assistance. The team is further indebted to Jeff Keister, Jen Bohon, Triveni Rao, Dimitre Dimitrov and Ilan Ben-Zvi for discussion and guidance over the course of this work. The authors wish to acknowledge the support of the U.S. Department of Energy under grant DE-FG02-08ER41547. This manuscript has been authored by Brookhaven Science Associates, LLC under Contract No. DE-AC02-98CH10886 with the U.S. Department of Energy.

REFERENCES

1. C. Bauer, I. Baumann, C. Colledani, J. Conway, P. Delpierre, F. Djama, et. al., Methods Phys. Res. Sect. A-Accel. Spectrom. Dect. Assoc. Equip. 383, 64 (1996).
2. M. J. Guerrero, D. Tromson, R. Barrett, R. T. Tachoueres, and P. Bergonzo, Physica Status Solidi a-Applied Research 201, 2529 (2004).
3. M. P. Gaukroger, P. M. Martineau, M. J. Crowder, I. Friel, S. D. Williams, and D. J. Twitchen, Diamond and Related Materials 17, 262 (2008).
4. J. W. Keister and J. Smedley, Nuclear Instruments & Methods in Physics Research Section A-Accelerators Spectrometers Detectors and Associated Equipment 606, 774 (2009).
5. C. Szeles, SE. Cameron, JO. Ndap, WC. Chalmers, *IEEE Transactions on Nuclear Science*, **49(5)**, 2535-2540 (2002).
6. A. R. Lang, R. Vincent, N. C. Burton, *J. Appl. Cryst*, **28**, 690-699 (1995).
7. H. Klapper, *Mater. Chem. Phys.* **66**.101-109 (2000)
8. H. Pinto, R. Jones, J. Phys.: Condens. Matter **21**, 364220 (2009)
9. M. Dudley, X.R. Huang, W. Huang, A. Powell, S. Wang, P. Neudeck, and M. Skowronski, *Appl. Phys. Lett.*, 75, p. 784-786, (1999).
10. Y. Mori, H. Kawarada, and A. Hiraki, *Appl. Phys. Lett.*, 58 (9), p. 940-941 (1991).

Mater. Res. Soc. Symp. Proc. Vol. 1203 © 2010 Materials Research Society 1203-J17-10

Deep UV Photodetectors Fabricated from CVD Single Crystal Diamond

Mose Bevilacqua and Richard B. Jackman
London Centre for Nanotechnology and Department of Electronic Engineering, University College London, 17-19 Gordon Street, London, WC1H 0AH, UK

ABSTRACT

Deep UV detection using a single crystal diamond (SCD) substrate without a homoepitaxial layer has been demonstrated using a defect passivation treatment. Despite evidence of surface damage on the SCD, the treatments lead to highly effective photoconductive devices, displaying six-orders of discrimination between deep UV and visible light and a responsivity as high as 100A/W, equivalent to an external quantum efficiency of 700, similar to the best values for devices based on high quality homoepitaxial layers. Impedance spectroscopic investigations suggest that the treatment used reduces the impact of less resistive surface material, most likely defects left from substrate polishing.

INTRODUCTION

It has long been considered that diamond, with a band gap of 5.5eV (225nm), is an ideal material for the fabrication of high performance deep UV photodetection devices. Landstrass and co-workers[1] investigated homoepitaxial single crystal layers grown by chemical vapour deposition (CVD) methods onto natural diamond substrates. The insulating properties of diamond suggest that simple photoconductive devices can be constructed which will display low dark currents without the need for reverse-biased p-n junctions. Devices with an interdigitated planar electrode configuration (35μm gaps, 10μm electrode widths) illuminated with pulsed 6.1eV light were used to determine that photoconductive mobilities as high as 3500 cm^2/Vs could be measured in the epitaxial layer, comparable to the best type IIa natural gem stones. No spectral response measurements were reported. More recently, Liao and Koide[2] fabricated photoconductive devices using an interdigitated planar metal-semiconductor-metal (MSM) structure on epitaxial layers grown by CVD on an HPHT type Ib substrate (10μm spacing, 10μm electrodes). Excellent spectral resolution was achieved, being 10^8 between deep UV (220nm) with a responsivity of around 6A/W, which is equivalent to an external quantum efficiency (EQE or gain) of 33. Whilst very useful, high quality epitaxial layers must be relatively thick to reduce dislocations densities that arise when the layer is grown on a polished substrate, making any device fabricated this way costly to produce. In terms of the realisation of a commercially useful device, polycrystalline CVD grown diamond films (PCD) have been investigated by McKeag et al[3] using an interdigitated planar MSM configuration with 20μm gaps and 20μm metal electrodes. Whilst the defective polycrystalline diamond meant that 'as-fabricated' photoconductive structures performed poorly, the use of post-growth defect passivation treatments dramatically improved device performance. Discrimination between deep UV light (220nm) and visible wavelengths exceeded 10^6, ultra-low dark currents were observed and a deep UV sensitivity of around 0.6A/W was reported. Similar performance levels with PCD without the use of similar defect passivation treatments were not acheived[4-10]. The work of McKeag[3] and co-workers was followed up by further enhancements in the use of defect passivation treatments to realise external quantum efficiencies (EQE, or gain) as high as 10^5 [11], to increase device speed for KHz operation[12] and for use within imaging systems[13]. Single crystal diamond (SCD) grown by CVD methods has recently become a commercially available material[14], awakening interest in the use of this form of diamond for the fabrication of UV photodetectors rather than PCD. Despite an extensive literature on the use of homoepitaxial single crystal layers grown on natural or HPHT substrates as photoconductive devices[1,2,15-19], the direct use of a CVD grown single crystal has attracted little attention. This letter describes the fabrication of devices made on commercially supplied SCD, and shows how the defect passivation treatments discussed above can be used to modify the surface of this form of SCD resulting in extreme levels of sensitivity to deep UV whilst maintaining good UV to visible light discrimination and dark current levels.

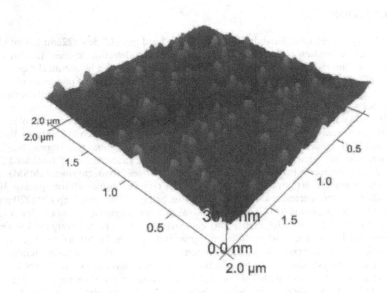

Figure 1
Atomic force microscope (AFM) typical of the cleaned single crystal diamond substrates (note Z-scale is 0-30nm)

EXPERIMENTAL METHODS

SCD substrates (produced by Element Six Ltd and supplied by Diamond Detectors Ltd) were subjected to wet treatments that are known to leave the surface free from contamination[20] and in an oxidised state avoiding so-called 'surface conductivity'[21]. Atomic force microscopy (AFM, Veeco Dimension V) was carried out, a typical image is shown in figure 1. The flat region of the substrate can be seen to be grooved with a periodicity of 0.05-$0.1\mu m$, regular protrusions from the surface are also apparent with a vertical dimension of around 10nm (note exaggerated Z-scale). The roughness of the surface can be found to have an rms value of around 3.5nm, although this drops to around 0.3nm when the measurement avoids the protrusions. Standard lithographic methods were used to pattern the metal electrodes (15nm Cr, 300nm Au), into an interdigitated array of 15 electrode pairs, spaced by $25\mu m$ on a $25\mu m$ pitch. Contact pads connected to each electrode pair were $300 \times 300\mu m$ in size. Defect passivation was performed in a way described previously[3], which involves treating the fabricated devices in a methane environment at 700°C for ca. 15 mins, followed by annealing in air at 400°C for one hour. To record the photoconductive response of the devices, the contact pads were probed and connected to a Keithley voltage source – picoammeter (model 487). Illumination from a monochromated Xe lamp source with dual gratings enabled 180-800nm wavelength light to be selected (Amko A1020). Impedance Spectroscopy (IS) was performed over the frequency range 0.1Hz-10MHz (Solartron 1260 Impedance system with Solartron 1296 Dielectric Interface). All experiments were carried out inside a stainless steel vacuum chamber (base pressure $\sim 10^{-3}$ mbar), which allowed both electrical screening and control of the ambient gas.

RESULTS

The spectral response of both 'as-fabricated' and passivated devices are shown in figure 2(a), where responsivity (A/W) is plotted as a function of illumination wavelength, for a device bias of 30V. In the case of the 'as-fabricated' device the peak responsivity at 210nm is around 0.1A/W, with the visible response being some six orders less, in the noise of the measurement system. Whilst there is a sharp decline in responsivity at the band edge (4 orders), some photoconductive signal persists until around 450nm. In contrast, the treated device shows a peak responsivity of ca. 10A/W, whilst retaining six orders of magnitude discrimination between the deep UV and visible light; however, the visible responsivity value remains above noise levels (at $\sim 10^{-4}$A/W). Figure 2(b) shows the 210nm responsivity value plotted against device bias level (left axis) and the same data expressed as external quantum efficiency (EQE or Gain, right axis). The device displays a responsivity of some 100 A/W at a bias level of 90V, equivalent to an EQE of around 700. Across this measurement range the background dark current remained in the nA scale range.

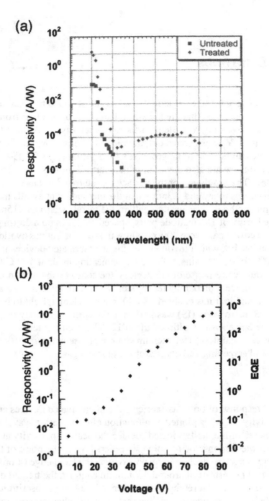

Figure 2
(a) Responsivity (A/W) plotted as a function of device illumination wavelength for untreated and treated single crystal photoconductive devices, biased at 30V (b) Responsivity (A/W) plotted as a function of device bias (V), left axis, and converted to external quantum efficiency (EQE), right axis

Impedance spectroscopy has been used previously to isolate different contributions to the overall conductivity of diamond samples[22-24], and was used here to give some insight into the changes being imparted to the SCD by use of the passivation treatment. The real and imaginary components of the impedance value were measured as a function of frequency, and plotted against each other in a so-called 'Cole-Cole' plot (figure 3(a)). The presence of a semicircular response indicates the presence of a conduction path within the sample being analysed with characteristic values of resistance and capacitance (R and C in parallel); observation of multiple semicircles arises through the presence of several conduction paths[25]. Figure 3(a) shows the Cole-Cole plots for both treated and untreated devices. Whilst only the initial part of a semicircle is apparent for the untreated device, a complete semicircle can be seen for the treated structure. The difference can be seen to arise from the different impedance values, and mathematical fitting to these curves gives R and C values of $8 \times 10^{11}\Omega$ and 0.5pF (untreated) and $1 \times 10^{11}\Omega$ and 3pF (treated); the fitting process suggests that these values may have an error of up to 10%. Figure 3(b) shows similar data measured at higher frequencies; whilst the plot for the treated device appears unchanged, a second semicircular response can be determined from the shoulder to the main feature on the untreated device. Mathematical fitting (shown) gives R and C values of $4 \times 10^9\Omega$ and 0.3pF respectively.

DISCUSSION

The changes in the spectral response curve shape and absolute responsivity following the use of the passivation treatment are marked (figure 2(a)). The untreated device shows some responsivity to wavelengths as long as 450nm; the shape of the spectral response curve is similar to that reported by Baducci and co-workers[19] when investigating the photoconductive response of high-quality CVD grown thick (150μm) homoepitaxial layers. This form of sub-band gap response indicates the presence of defects and trap states within the material. In contrast, the treated device shows a much sharper decline in responsivity at wavelengths greater than 220nm, although some photoconduction persists at all measured wavelengths, indicative of non-diamond carbon. The measured value of EQE (fig. 2(b)) is comparable for a given field strength to that recorded by Liao and Koide[3] working with high quality homoepitaxial diamond films, and considerably higher than other reports using homoepitaxial films[15,19]. This indicates that the passivation treatment affects changes to the single crystal substrate making it at least as good as the very best homoepitaxial films. The mobility-lifetime product, μτ, can be evaluated from the following expression for a photoconductor[26]

$$G = \mu\tau V/L^2 \quad [1]$$

where G represents gain, also known as the external quantum efficiency (EQE), V is the applied voltage and L the gap between the electrodes. In the current case the μτ product can be calculated to be around 5×10^{-5} cm^{2}V^{-1}. This value is similar to that found by Liao and Koide[3], further highlighting the high quality, in terms of carrier generation and transport, of the SCD substrates used here following treatment. The penetration depth of 210nm light in diamond is of the order of 3μm[27], suggesting that the passivation treatment used here results in material changes potentially within this depth.

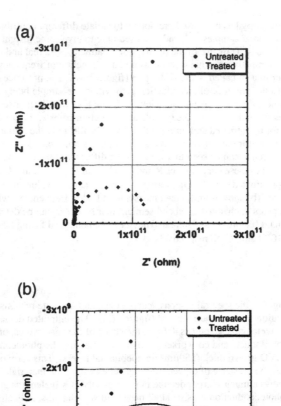

Figure 3
(a) Cole-Cole plot (real impedance, Z' vs imaginary component of impedance, Z'') for the treated and untreated device (b) similar data to that plotted in (a) but recorded at high frequencies. The solid line is a mathematical fit to the semicircular response represented by the shoulder in the data for the untreated device.

The observation of two semicircular responses in the IS measurements on the untreated SCD devices (figure 3) is indicative of two conduction paths within the material. In polycrystalline materials is well established that pF range capacitance levels are associated with grain interior conduction, where as nF range capacitance is suggestive of grain boundary conduction[25]. The capacitance levels for the two processes measured for the untreated device are both in the pF regime, as to be expected for SCD; they principally differ in terms of the resistivity associated with each, which changes by around two orders of magnitude. In the case of the treated device, the IS component associated with the higher resistivity in the untreated case remains, whilst the lower resistivity component has been removed. The most likely explanation is that the treatment eradicates disordered material in the surface region that is responsible for the lower resistivity conduction process. Given that the treatment considerably enhances the EQE, and hence the μτ product, is can be inferred that this region is responsible for the relatively poor characteristics displayed by the 'as-fabricated' untreated devices. Previous work carried out in the authors laboratories suggested that the passivation treatment used here, worked on polycrystalline materials by diffusing carbon atoms into voids and vacancies that otherwise act as trapping and recombination centres[28]. It would seem likely that in the current case, structural damage left by substrate polishing processes (fig.1) are being passivated in a similar manner, giving rise to improved values of the μτ product.

CONCLUDING REMARKS

Previously developed treatments that passivated defects in PCD material, leading to enhanced deep UV photoconductive device performance have been found to act similarly to improve SCD devices, formed from commercially purchased diamond. The quality, in morphological and electrical terms, of PCD varies significantly between laboratories and commercial suppliers. In contrast high quality SCD substrates are a more stable platform for device fabrication. Reports on the use of such material have been encouraging, but have involved the growth of a high quality homepitaxial film on the substrate prior to device fabrication. In the current study extreme sensitivity to deep UV light has been achieved without the need for this process, by application of a simple defect passivation treatment. This should enable relatively low cost high performance diamond UV photodetection devices to be realized in the near future.

ACKNOWLEDGEMENTS

The authors are indebited to Dr Robert McKeag of Centronic Ltd for useful discussions and practical advice. Dr Steve Hudziak (UCL) is also thanked for performing the AFM measurements presented here.

REFERENCES

1. M. I. Landstrass, M. A. Plano, M. A. Moreno and S. McWilliams, L. S. Pan, D. R. Kania and S. Han Diamond and Related Materials, 2, 1033 (1993)
2. M. Liao and Y. Koide, Appl. Phys. Letts., 89, 113509 (2006)
3. R. D. McKeag, S. S. M. Chan, and R. B. Jackman, Appl. Phys. Letts., 67, 2117 (1995)
4. S. C. Binari, M. Marchywka, D. A. Koolbeck, H. B. Dietrich, and D. Moses, Diam. Relat. Mater. 2, 1020 (1993)
5. S. Salvatori, R. Vincenzoni, M.C. Rossi, F. Galluzzi, F. Pinzari, E. Cappelli, P. Ascarelli, Diamond and Related Materials 5, 775 (1996)
6. S. Salvatori, E. Pace, M.C. Rossi, F. Galluzzi, Diamond and Related Materials 6, 361 (1997)
7. S. Salvatori, M.C. Rossi, F. Galluzzi, E. Pace, P. Ascarelli, M. Marinelli
   Diamond and Related Materials 7, 811 (1998)
8. V.I. Polyakov, A. Rukovishnikov, N.M. Rossukanyi, A.I. Krikunov, V.G. Ralchenko, A.A. Smolin, V.I. Konov, V.P. Varnin, I.G. Teremetskay
   Diamond and Related Materials 7, 821 (1998)
9. S. Salvatori, M.C. Rossi, D. Riedel, M.C. Castex
   Diamond Relat. Mater. 8, 871 (1999)
10. E. Lefeuvre, J. Achard , M.C. Castex , H. Schneider , C. Beuille , A. Tardieua
    Diamond and Related Materials 12, 642 (2003)
11. R. D. McKeag and R. B. Jackman
    Diamond and Related Materials 7, 513 (1998)
12. S. P. Lansley, O. Gaudin, M. D. Whitfield, R. D. McKeag, N. Rizvi and R. B. Jackman, Diamond and Related Materials 9, 195 (2000)
13. S. P. Lansley , O. Gaudin , H. Ye , N. Rizvi , M. D. Whitfield, R. D. McKeag , R. B. Jackman Diamond and Related Materials 11, 433 (2002)
14. See, for example, www.e6cvd.com
15. R. Brescia, A. De Sio, M. G. Donato, G. Faggio, G. Messina, E. Pace, G. Pucella, S. Santangelo, H. Sternschulte, and G. Verona Rinati
    phys. stat. sol. (a) 199, 113 (2003)
16. T. Teraji, S. Yoshizaki, H. Wada, M. Hamada, T. Ito
    Diamond and Related Materials 13, 858 (2004)
17. A. Balducci, A. De Sio , Marco Marinelli, E. Milani, M.E. Morgada, E. Pace, G. Prestopino, G. Pucella, M. Scoccia, A. Tucciarone, G. Verona-Rinatia
    Diamond & Related Materials 14, 1980 (2005)
18. A. Balducci, Marco Marinelli,a! E. Milani, M. E. Morgada, A. Tucciarone, and G. Verona-Rinati, M. Angelone and M. Pillon
    Appl. Phys. Letts., 86, 193509 (2005)
19. Y. Iwakaji, M. Kanasugi, O. Maida and T. Ito
    Appl. Phys. Letts., 94, 223511 (2009)
20. B Baral, SSM Chan and RB Jackman
    J. Vac. Sci. & Technol., A14, 2303 (1996)
21. OA Williams and RB Jackman
    J. Appl. Phys., 96, 3742 (2004)
22. H. Ye, O. A. Williams, R. B. Jackman, R. Rudkin, and A. Atkinson
    Phys. Status Solidi A-Appl. Res. 193, 462 (2002)
23. HT Ye, RB Jackman, P Hing, J. Appl. Phys., 94, 7878 (2003)

24. S Curat, H Ye, O Gaudin, RB Jackman, S Koizumi
     J. Appl. Phys., **98**, 073701 (2005)
25. L. L. Hench and J. K. West, Principles of Electronic Ceramics (Wiley, New York, 1989), Chapter 5
26. R. H. Bube, *Photoelectronic Properties of Semiconductors* (Cambridge, New York, 1992, p. 32)
27. R. Brescia, A. De Sio, E. Pace and M.C. Castex
     Diamond and Related Materials 13, 938 (2004)
28. O. Gaudin, S. Watson, S. P. Lansley, H.J. Looi, M. D. Whitfield and R. B. Jackman
     Diamond and Related Materials 8, 886 (1999)

Mater. Res. Soc. Symp. Proc. Vol. 1203 © 2010 Materials Research Society   1203-J17-21

Diamond Photodiodes for X-ray Applications.

J. Smedley[1]; J. W. Keister[1]; E. Muller[3]; J. Jordan-Sweet[5]; J. Bohon[6]; J. Distel[2]; B. Dong[4]
[1]Brookhaven National Laboratory, Upton, NY, USA.
[2]Los Alamos National Laboratory, Los Alamos, NM, USA.
[3]Stony Brook University, Stony Brook, NY, USA.
[4]Global Strategies Group, North America, Crofton, MD, USA.
[5]IBM T.J. Watson Research Center, Yorktown Heights, NY, USA.
[6]Center for Synchrotron Biosciences, Case Western Reserve University, Upton, NY, USA.

ABSTRACT

Single crystal high purity CVD diamonds have been metallized and calibrated as photodiodes at the National Synchrotron Light Source (NSLS). Current mode responsivity measurements have been made over a wide range (0.2-28 keV) of photon energies across several beamlines. Linear response has been achieved over ten orders of magnitude of incident flux, along with uniform spatial response. A simple model of responsivity has been used to describe the results, yielding a value of 13.3±0.5 eV for the mean pair creation energy. The responsivity vs. photon energy data show a dip for photon energies near the carbon edge (284 eV), indicating incomplete charge collection for carriers created less than one micron from the metallized layer.

INTRODUCTION

Diamond is an attractive material for x-ray windows and transmission x-ray monitors due to its low Z and good thermal properties. Diamond also has application in high radiation environments as an x-ray sensor [1-4]. However, only recently has material become commercially available with low enough impurity and defect density to realize its potential in radiometric applications such as synchrotron diagnostics and scientific instrumentation [5, 6]. The present report describes the observation of calculable diode responsivity under controlled conditions, as measured at several NSLS synchrotron beamlines, spanning a wide range of photon energies and flux levels. Methods of metallization and surface preparation will be discussed, along with the effect of thermal annealing on these contacts.

METALLIZATION and SURFACE PREPARATION

Several methods of creating electrical contacts on diamond have been investigated. These can be broken into three broad categories – metal contacts on oxygen-terminated diamonds, thermally annealed contacts using carbide-forming metals, and annealed contacts using non-carbide forming metals. The process used to create each of these contacts will be described in this section, along with physical analysis of the contacts.

All of the devices discussed herein begin with single crystal "detector grade" material obtained from Element6. These plates are typically 4x4 mm^2 and 0.25-0.5 mm thick with a [001] surface orientation and a [110] edge normal. The surface finish has been measured with atomic force microscopy and found to be consistent with the manufacturer's specification of < 4 nm roughness. Prior to metallization, the plates are treated with a standard chromic acid etch to both remove surface non-diamond carbon and leave the surface oxygen terminated. Subsequent oxygen termination has also been achieved via exposure to an ozone lamp for several hours. Near-edge x-ray absorption fine structure (NEXAFS) oxygen-edge analysis has verified that both methods are capable of producing an oxygen-terminated surface.

Metal contacts, typically ~30 nm thick, are formed via sputtering. A mask is used to create a 3 mm diameter contact on the center of both sides of the diamond. Several metals have been investigated on oxygen-terminated diamond – copper, platinum, niobium, molybdenum, and titanium (with a platinum capping layer). In all cases, prior to thermal annealing, the contacts were found to be blocking – meaning that they did not support photoconductive gain/persistent photocurrent, and there exists a bias beyond which the response is independent of field.

Thermal annealing was used on some samples to demonstrate the creation and properties of carbide contacts and to prepare detectors which were intended to operate in high flux (and resulting high temperature) environments. NEXAFS analysis indicates that temperatures above 600 °C cause loss of oxygen termination; this leads to injecting contacts for all metals tested, however the extent and location of the resulting photoconductive gain varies greatly with contact. For high-flux x-ray irradiation, it is the temperature that the detector reaches, and not the irradiation, which causes the loss of oxygen and subsequent change in the contact nature. To investigate this, a sample was irradiated with focused white beam at NSLS beamline X28C (100 W/cm^2 with an energy range of 5-15 keV) for 1 min; oxygen edge NEXAFS showed no change in the oxygen-edge feature due to this irradiation.

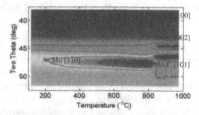

Figure 1 - Annealing of Mo contact to form Mo$_2$C. Color scale provides peak intensity.

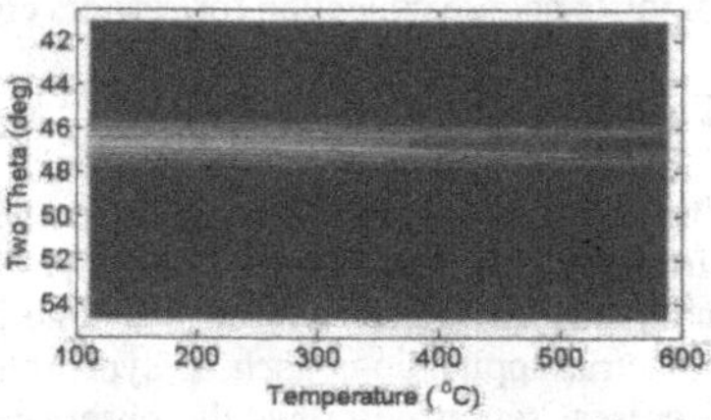

Figure 2 - Annealing of Pt contact showing [111] texture growth.

Annealing was performed in a He-filled oven with *in-situ* x-ray diffraction at beamline X20C. For carbide forming metals, this capability allows the formation of the carbide to be directly observed and the transition temperature to be measured. Figure 2 shows the formation of Mo$_2$C contacts on diamond, starting from a 30 nm Mo film. The ramp rate is 3 °C/s, and the transition temperature is 835 °C. For Pt contacts intended for high flux applications, annealing in the x-ray diffraction (XRD) oven reveals a significant increase in grain size above 400 °C. This film has a [111] fiber texture which strengthens upon heating. The peaks become significantly narrower during annealing, indicating an increase in grain size. Subsequent Scherrer analysis of the diffracting grain size via high-resolution XRD at X20A revealed that the pre-anneal [111] fiber-textured grain size was ~50 Å, and the post anneal size was ~200 Å. Other random texture components had smaller grain sizes, down to 26 (pre-anneal) and 70 (post-anneal) Å.

DETECTOR CALIBRATION

Calibration of detectors has been performed across five NSLS beamlines – U3C, X6B, X8A, X15A and X28C. Together, these sources provide photon energies from 200 eV to 28 keV and flux from 100 pW to 10 W, with beam diameters from 20 µm to 2 mm. Calibration is done via comparison to a silicon photodiode at low flux and soft/tender x-rays, an ion chamber at mid flux, and a copper calorimeter at the highest flux. One side of a detector is biased, and the other is used for charge collection. Both pulsed and DC biases are used; here pulsed bias is a square wave defined by a duty cycle and frequency as well as amplitude and polarity. Current through the detector is monitored using Keithley electrometers (models 617 & 6517). For the soft x-ray lines (U3C and X8A), the detectors are calibrated in vacuum; all other calibrations are performed in ambient conditions, with the exception of high flux testing at X28C, where a dry N_2 environment is used to limit ozone production.

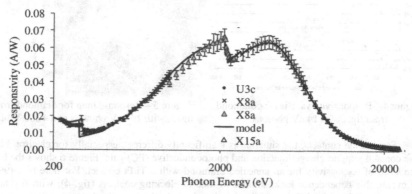

Figure 3 – Responsivity vs. photon energy for a 260 μm thick oxygen-terminated diamond with 40 nm thick platinum contacts. 200 V bias on incident electrode was used (hole bias), with 1 kHz, 95% duty cycle for energies up to 1 keV and 100% duty cycle beyond 1 keV.

Figure 3 shows the responsivity of a diamond with blocking platinum contacts (on oxygen terminated diamond) as a function of incident photon energy for positive bias on the incident electrode. In the soft x-ray regime, we refer to this as hole mode, as the current is carried primarily by holes pushed away from the incident electrode and collected on the transmission side. Beyond 4 keV, this distinction no longer holds, as the photons are absorbed (and thus carriers are created) throughout the thickness of the diamond. Note that the flux of the various beamlines used to create figure 3 varies widely with photon energy; thus the fact that the responsivity follows a predictable theory (see discussion for details on the model) implies that the diamond response is linear in flux over the range of 100 pW to 10 μW. High flux measurements have extended this linearity up to 1 W [7], but under these conditions the blocking nature of the contact is lost due to heating. Below 1 keV photons, a pulsed bias is used to enable full collection. This is necessary to prevent the buildup of trapped charge; carriers produced by the beam during the "off" portion of the cycle drift under the influence of the trapped charge, eventually neutralizing it. For hole bias, full collection can be achieved with up to 99% duty cycle at 0.1 MV/m (25 V across a 0.25 mm thick diamond). For electron transport, trapping is a more significant problem, and higher fields or smaller duty cycles are required. Figure 4 shows the responsivity vs. bias for both polarities with 1 keV photons (the bias is on the incident electrode, thus the "hole" bias is positive and the "electron" bias is negative). The spatial uniformity of detectors with blocking contacts is excellent – figure 5 shows a spatial response map with 19 keV photons taken at X15A under +200 V bias. The circular area corresponds to the metalized region of the device, while the notch at the bottom comes from the mount used to hold the diamond. The x-ray beam used for the map is 0.2x0.2 mm². The area of increased response near the mount is an artifact of x-rays striking the copper holder. The scale is referenced to the expected response of 2.2 mA/W at this photon energy.

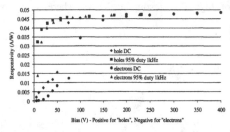

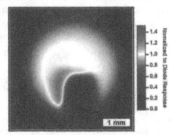

Figure 4 – Responsivity vs. Bias for diamond from fig. 3, for 1 keV photons.

Figure 5 – Response map for diamond from fig. 3, with 19 keV photons and +200 V bias.

For carbide contacts, the situation is significantly different, especially for electron bias. These contacts support charge injection and photoconductive (PC) gain. Figure 6 shows the bias dependence of responsivity for an annealed diamond with a Ti/Pt contact. For hole bias (99% duty cycle), this dependence looks similar to that of blocking contacts (fig. 4), with full hole collection at a field of 0.1 MV/m - indicating that the hole charge collection distance is equal to the thickness of the diamond at this field. The electron transport for low fields is significantly less than the expected value, indicating charge trapping. For fields > 0.1 MV/m, the responsivity is significantly *greater* than expected, because the trapped electrons lead to hole injection (thus increasing the effective hole lifetime). These injected holes cross the diamond many times before the electrons are detrapped (or the holes are themselves trapped). This can only occur for field at which the hole charge collection distance is greater than the thickness of the material (this explains why there is no gain below 0.1 MV/m). Above this value, the gain should be roughly linear with bias, as the hole velocity is linear with bias in this regime. The responsivity vs. photon energy for a 0.3 mm thick diamond with annealed Ti/Pt contacts is shown in figure 7 for three bias conditions.

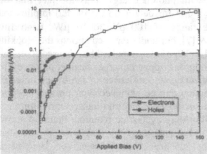

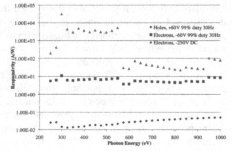

Figure 6 – Responsivity vs bias for a diamond with annealed Ti/Pt contacts; 1 keV photons.

Figure 7 – Responsivity vs. photon energy for diamond from fig. 6.

The beamline flux increases by two orders of magnitude from 550 eV to 600 eV. This change is evident in the electron response, especially for DC bias, demonstrating that PC gain is non-linear with flux. For hole bias, however, the response is similar to the blocking case (fig. 3), again suggesting that electron trapping is the cause of the PC gain. For carbide contacts, gain occurs everywhere that electrons are incident on the contact. Figure 8 shows response maps of a diamond with an injecting Mo_2C contact on one side and an injecting Mo contact on the other side. These maps were taken at 19 keV, so both carriers are moving throughout the material. PC gain is observed at all points when the bias is set to move electrons toward the Mo_2C contact - every point in the metalized area is at least 5 times the expected response; some areas are much higher. For the opposite bias, the gain is much lower, and is observed only in isolated areas (similar to the annealed platinum discussed below).

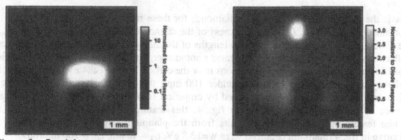

Figure 8 – Spatial response maps for a diamond with one carbide contact. Note the difference in color scale. The map on the left, with the diamond biased (-200 V DC) to move electrons toward the Mo_2C contact, shows gain over the entire metalized region. The map on the right, with the opposite bias (+200 V) shows significantly lower and more isolated PC gain.

Annealed platinum contacts are also injecting, but PC gain occurs only at locations where defects appear in x-ray topographs of the material (specifically single-point strain fields associated with inclusions). In contrast to the carbide contacts, these areas of gain are typically only ~100 um across, and some diamonds with low defect density show very few of these spots.

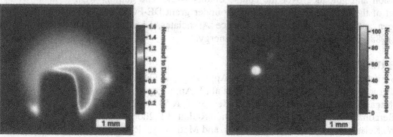

Figure 9 – Spatial response maps for a diamond with annealed platinum contacts, for both -200V (left) and +200 V (right) bias. PC gain in observed only in two locations, and only for + bias.

Figure 9 shows response maps of a diamond with annealed Pt contacts under positive and negative bias at 19 keV. PC gain occurs at different locations for each polarity, leading us to conclude that the feature responsible for the charge trapping must be near the hole injecting electrode to cause gain. Away from these defect regions, the annealed Pt contacts behave similarly to the blocking contacts, providing predictable, diode like response. Once the contact is annealed, its behavior remains consistent at any flux level – diamonds with annealed Pt contacts have been shown to be linear with a dynamic range of *11 orders of magnitude*.

DISCUSSION and CONCLUSIONS

The responsivity (S) of the diamonds with blocking contacts (or annealed contacts away from PC regions) is predictable by a simple expression:

$$S[\nu] = (\frac{1}{w})(e^{-\frac{t_{metal}}{\lambda_{metal}[\nu]}})(1 - e^{-\frac{t_{dia}}{\lambda_{dia}[\nu]}})CE[\nu, F]$$

Here w is the mean ionization energy of diamond; for these measurements, 13.25 ± 0.5 eV has been obtained [5]. t_{dia} and t_{metal} are the thickness of the diamond and the metal, and λ_{dia} and λ_{metal} are the photon-energy dependent absorption lengths of the diamond and the metal. CE represents the collection efficiency of the device. For most photon energies, CE is unity as long as the field (F) is sufficient for full collection. For photons near the carbon K edge (284 eV), the carriers are created very near the incident electrode (under 100 nm); this leads to loss of carriers due to diffusion into the metal layer. CE is obtained by empirical fit to monte carlo results [8]. This is the source of the carbon edge feature in fig. 4; this feature is bias dependent, unlike the absorption feature at 2120 eV which results from the platinum M edge. The parameters for model curve in fig. 4 (platinum contacts) were w=13.8 eV, t_{Pt} = 40 nm, t_{dia} = 260 microns.

We have demonstrated that diamond can be used to create calibrated photodiodes for x-ray applications. It is important to both choose appropriate contact materials for a given environment and to be aware that pulsed biasing may be required to achieve full carrier collection, particularly for soft x-ray applications.

ACKNOWLEDGMENTS

The authors wish to thank John Walsh for design and fabrication of sample mounts and Xiangyun Chang for assistance with metallization. The team is further indebted to Veljko Radeka, Pavel Rehak, Peter Siddons, Triveni Rao, Dimitre Dimitrov and Ilan Ben-Zvi for discussion and guidance over the course of this work. The authors wish to acknowledge the support of the U.S. Department of Energy under grant DE-FG02-08ER41547. This manuscript has been authored by Brookhaven Science Associates, LLC under Contract No. DE-AC02-98CH10886 with the U.S. Department of Energy.

REFERENCES

1. D. R. Kania, L. S. Pan, P. Bell, et al., J. Appl. Phys. 68 (1) (1990) 124.
2. H. Pernegger, S. Roe, P. Weilhammer, et al., J. Appl. Phys. 97 (7) (2005) 073704.
3. H. Kagan, Nucl. Instr. and Meth. Phys. Res. Sect. A 541 (1-2) (2005) 221.
4. P. Bergonzo, D. Tromson, C. Mer, J. Synch Radiat. 13 (2006) 151.
5. J. W. Keister and J. Smedley, Nucl. Instr. and Meth. Phys. Res. Sect A 606 (2009) 774.
6. J. Morse et al., Mater. Res. Soc. Symp. Proc. 1039 (2008) P06-02.
7. J. Bohon et al., these proceedings.
8. D. Dimitrov et al., these proceedings

Mater. Res. Soc. Symp. Proc. Vol. 1203 © 2010 Materials Research Society 1203-J19-03

Development of Diamond-Based X-ray Detection for High Flux Beamline Diagnostics

Jen Bohon[1], John Smedley[2*], Erik Muller[3] and Jeffrey W. Keister[4]

[1] Center for Synchrotron Biosciences, Case Western Reserve University, Upton, NY 11973, USA.

[2] Instrumentation Division, Brookhaven National Laboratory, Upton, NY 11973, USA.

[3] Stony Brook University, Stony Brook, NY 11794, USA.

[4] NSLS-II, Brookhaven National Laboratory, Upton, NY 11973, USA.

ABSTRACT

High quality single crystal and polycrystalline CVD diamond detectors with platinum contacts have been tested at the white beam X28C beamline at the National Synchrotron Light Source under high-flux conditions. The voltage dependence of these devices has been measured under DC and pulsed-bias conditions, establishing the presence or absence of photoconductive gain in each device. Linear response has been achieved over eleven orders of magnitude when combined with previous low flux studies. Temporal measurements with single crystal diamond detectors have resolved the ns scale pulse structure of the NSLS.

INTRODUCTION

Measurement of monochromatic synchrotron x-ray beam flux is traditionally performed using gas-filled ion chambers and silicon photodiodes. These devices saturate under high-flux conditions, however, necessitating the use of alternative tools. The X28C beamline at the National Synchrotron Light Source (NSLS) produces a focused x-ray white beam capable of delivering nearly 90 W/mm^2 to the focal point. Diagnostics for this beam are problematic not only due to saturating levels of flux, but also due to significant thermal effects that occur at these power densities. Many state-of-the-art synchrotron beamlines face similar challenges. Diamond is a material particularly well suited to addressing these issues,[1-3] as it has a unique combination of mechanical, optical, electronic and thermal properties. While diamond is currently used for white-beam diagnostics, it has typically been for its thermal, crystalline or transmission properties. Here we describe a diamond-based photodiode, which also exploits the electronic properties. Two types of diamond detectors (polycrystalline and single crystal) were tested at X28C under a variety of conditions. A linear response was obtained for all fluxes measured; this spanned eleven orders of magnitude (from 0.1 nW to 10 W), including measurements made at both monochromatic and white-beam beamlines. The diamond was exposed to absorbed x-ray power on the order of a Watt; the incident x-ray flux produced 40 mA of current in the 2 mm^2 area exposed to the beam. In addition, a temporal response was measured which clearly resolved the pulse structure of the NSLS, with indications that higher biasing and thinner diamond material might provide the capability to measure even shorter time scale signals.

EXPERIMENTAL SETUP

The focused white-beam X28C beamline at the NSLS is capable of producing a range of power densities and spot sizes up to a maximum of ~90 W/mm^2 in a focused 0.5 mm diameter spot with a broad energy range (5-15 keV).[4] This allows access to a range of broadband flux values through focusing of the mirror and the use of aluminum filters of various thicknesses (0.08-4.5 mm). A schematic of the experimental setup is shown in figure 1. Initial measurements were performed in air, causing significant damage to the metal contacts due to high ozone concentrations created by the x-ray beam, thus a dry nitrogen purged enclosure was built for subsequent experiments. Initially, the beampipe was kept at vacuum, however focusing of the mirror in later experiments caused enough heating of the terminal 0.076 mm aluminum window to melt a hole and break vacuum; later experiments were performed with a helium purge on the beampipe, leaving the pinhole in the window.

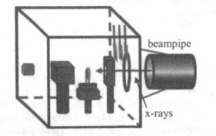

Figure 1. X28C Experimental setup. X-rays exit the beampipe, enter an N_2-filled enclosure and may pass through attenuators and a 1.6 mm diameter pinhole before striking the diamond. The beam then passes through an ion chamber and strikes a copper calorimeter.

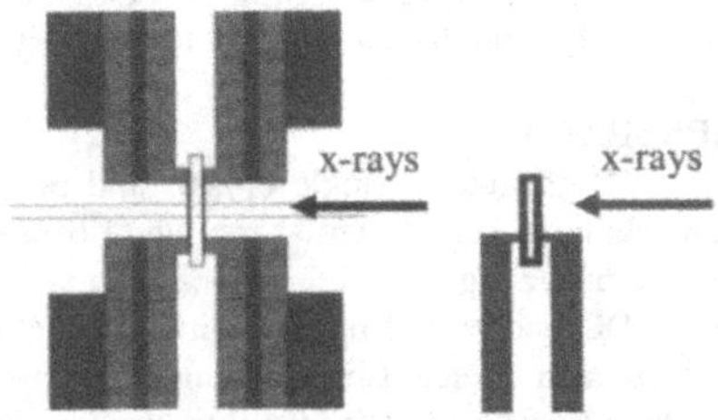

Figure 2. Diamond mounts. The holder on the left allows timing measurement, but holds only 4 mm square diamonds. The holder on the right can accommodate a range of diamond sizes and can be used for response mapping.

Two different mounting systems were used for the diamonds (Fig. 2). One mount was created to hold only the 4 mm square single crystal diamonds, with pressure contacts on either side and a rigid coaxial cable and bias-compensating capacitor capable of making timing measurements. The other mount has spring-loaded copper contacts which touch the diamond at a single point on either side of the diamond. This holder allows the use of a range of diamond sizes and can be used in a variety of measurements, including response mapping (performed at X6B), but cannot measure timing. All devices discussed have sputtered platinum contacts on both sides; the specifics of this process are described elsewhere.[5] Single crystal diamonds (4x4 mm^2, 0.25-0.5 mm thick) were metallized with a 3 mm diameter pattern and a polycrystalline diamond (1 cm diameter, 0.21 mm thick) were metallized with a 6 mm diameter pattern, centered on the diamond with non-metallized diamond extending beyond the contact to avoid flow of current around the edge of the device. Bias was applied to the incident surface of the diamond and current was measured from the opposite side using Keithley electrometers (models 617 & 6517). Both DC and pulsed voltages were applied for reasons which will be discussed below. Calibration of each device was performed via comparison to an ion chamber at lower fluxes, and a copper calorimeter at high flux.

RESULTS AND DISCUSSION
Detector Calibration
Both single crystal and polycrystalline devices were tested under high-flux conditions; the single crystal devices have also been calibrated at lower fluxes at monochromatic beamlines. The linearity of this calibration spans 11 orders of magnitudes over all of these measurements, six of which are shown in figure 3, performed at X28C. The low flux data collected at

monochromatic beamlines[6] yielded a mean ionization energy (W) of 13.3 eV which predicted well the response over a further 6 orders of magnitude for both single crystal and polycrystalline diamond (Fig. 3).

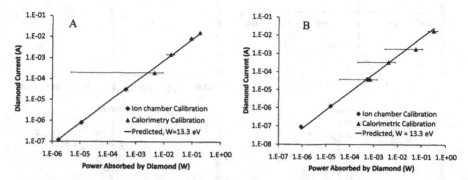

Figure 3. Calibration of A) Single Crystal (0.5 mm thickness, 300 V DC bias) and B) Polycrystalline (0.21 mm thickness, 200 V DC bias) diamond detectors.

Voltage Dependence

Figure 4 shows the dependence of the device response on voltage. In general, once full carrier collection has been achieved, the signal increase ends and the response becomes independent of voltage. For the single crystal diamond device shown in figure 4A, this occurs for positive biasing, but negative bias creates a linearly increasing signal with voltage after the initial phase – the hallmark of photoconductive (PC) gain. At the inflection point, the bias becomes high enough that the carriers injected due to trapped charge can move across the entire thickness of the diamond before they can be trapped or can neutralize the trapped charge.[7] As this injection can occur many times before the trapped charge is neutralized, we observe an associated gain in current. The addition of a pulsed-bias cleaning cycle (square wave pulse defined by a duty cycle, frequency, amplitude and polarity) allows detrapping of the material and is able to "turn off" the photoconductive gain; in this case (Fig. 4B), a 10% duty cycle was used for the negative polarity, while the positive bias was applied DC. Figure 4C depicts the response of a non-PC single crystal diamond, which exhibits diode-like behavior even under the highest flux conditions measured. X-ray topography and response mapping has indicated that this diamond is free of electrically active point defects in the active area.[8] This is not the case for the photoconductive diamond, as will be discussed later. Figure 4D shows the voltage dependence of a polycrystalline diamond. The characteristics of this curve are different in two ways from the single crystal case; the slope of the rise leading to saturation is lower, and the signal does not plateau, but continues to gradually rise for all voltages measured. The expected response value is obtained at the transition between these phases, suggesting that PC gain may play a role in the subsequent response increase with voltage.

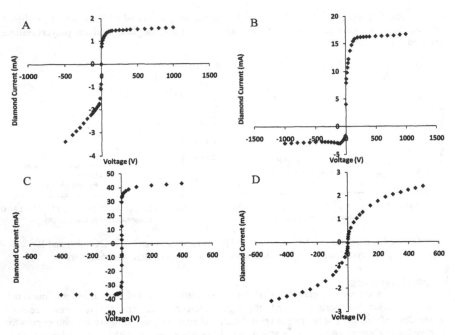

Figure 4. Voltage dependence of A) PC single crystal DC bias, B) PC single crystal DC + bias, pulsed - bias, C) non-PC single crystal DC bias and D) polycrystalline diamond DC bias.

Timing

Figure 5 shows timing measurements of the NSLS bunch structure. Of the 30 ~1 ns buckets available, 25 were filled, leaving 5 empty for ion cleaning (Fig. 5A). The observed pulse width is broadened due to the transit time of the carriers; the thickness of the diamond and the strength of the applied field will determine timing capabilities. The fast rise and slow fall behavior can be attributed to the creation of carriers throughout diamond, unlike a traditional time of flight measurement, which is done with α particles where only one carrier crosses the full

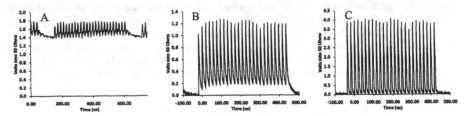

Figure 5. Timing measurements of A) photoconductive (PC) 100 V bias, B) non-PC 150 V bias 50% duty cycle and C) non-PC 1000 V DC bias single crystal diamonds.

thickness of the diamond. While the rise is limited by the ~ 1 ns pulse width, the decay is related to carrier transit. In the PC device (Fig. 5A), the decay is much longer due to the trap lifetime; in this particular case, we believe that the trapped carriers are electrons. The PC gain has reached a steady state of trapping and detrapping, thus the high steady baseline. The non-PC device under low biasing conditions (Fig. 5B) also illustrates the effect of carrier mobility on the timing resolution; here, the bias is not strong enough to drive the carriers across the diamond before the next incident pulse except during the ion cleaning cycle, creating an elevated baseline during the 25 bunch series. Under high voltage (Fig. 5C), the bias is able to move the carriers across in much less than the ~18.9 ns of the bunch period, bringing the signal back to zero between pulses. It is important to note that the current values calculated from these traces are also consistent with average current measurements on the electrometers for all devices characterized.

Lifetime

For high flux applications which require temporal resolution, considerable voltages are needed; high current draw through the diamond is the major cause of heating of the device. If timing measurements are not required, these types of detectors can be operated at low voltage (50 V), which will significantly decrease the heat load and simplify mounting and cooling requirements. The contact material used must be robust in the environment in which it will be used; early copper contacts evaporated when in the beam path, likely due to ozone produced in the air nearby; all subsequent high flux experiments have been performed in a nitrogen environment with platinum contacts to avoid this possibility. It is important to note that for high-flux measurements thus far, blocking contacts cannot be maintained. Once exposed to high-flux x-ray irradiation, the temperatures that the detectors reach cause the loss of oxygen from the diamond surface[5] and a subsequent change in the contact nature; basically the contacts are annealed in the x-ray beam.

Response mapping of the annealed contacts of the PC single crystal diamond clearly illustrate the isolated nature of the PC spots (Fig. 6). The negative bias response map indicates that there are two PC spots within the target area for the incident beam, while the only PC spot on the positive response map falls outside of that area and would therefore not be observable.

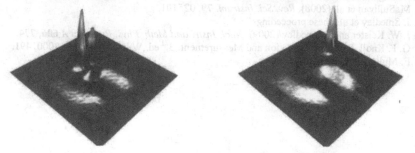

Figure 6. Response maps of PC single crystal diamond. Left: negative bias; right: positive bias. The rectangular pattern inset into the diamond response is the area occluded by the holder (data taken at X6B at 19 keV, 100 V bias in each polarity, 80% duty cycle).

These PC spots have been shown to correlate with point defects in the diamond using x-ray topography;[8] diamonds which do not contain these defects do not exhibit PC gain (Fig. 4C), indicating that improving the quality of the diamond within the detection region will result in diode-like behavior in devices constructed from this material.

CONCLUSIONS

We have demonstrated that CVD diamond detectors perform well over a wide range of flux values and can be used for temporal profiling of the beam. Both the quality and thickness of the diamond material used will determine the biasing requirements to achieve the desired measurement capabilities of a given device. For simple flux measurements using single crystal devices with no timing requirements, the bias can be as low as 50 V. The theoretical mean ionization energy can be used for flux calibration of both single crystal and polycrystalline devices. Ongoing studies include characterization of a four-channel quadrant version of the single crystal detector which was constructed for simultaneous flux and beam position measurements.

ACKNOWLEDGMENTS

The authors wish to thank John Walsh for design and fabrication of sample mounts, Xiangyun Chang for assistance with metallization and Bin Dong for technical assistance. The team is further indebted to Veljko Radeka and Triveni Rao for discussion and guidance over the course of this work. The authors wish to acknowledge the support of the U.S. Department of Energy (DOE) under grant DE-FG02-08ER41547 and the National Institute for Biomedical Imaging and Bioengineering under P30-EB-09998. This manuscript has been authored by Brookhaven Science Associates, LLC under Contract No. DE-AC02-98CH10886 with the U.S. DOE.

REFERENCES

1. J. Morse et al. (2008) *Mater. Res. Soc. Symp. Proc.* **1039**, P06-02.
2. P. Bergonzo et al. (2006) *J. Synch Radiat.* **13**, 151.
3. H. Kagan (2005) *Nucl. Instr. and Meth. Phys. Res. Sect. A* **541** (1-2), 221.
4. M. Sullivan et al. (2008), *Rev. Sci. Instrum.* **79**, 025101.
5. J. Smedley et al., these proceedings.
6. J. W. Keister and J. Smedley (2009), *Nucl. Instr. and Meth. Phys. Res. Sect A* **606**, 774.
7. G. F. Knoll, Radiation Detection and Measurement, 3rd ed.,Wiley, NewYork, 2000, 491.
8. E. Muller et al., these proceedings.

Device Applications

(4) Field Emission

Mater. Res. Soc. Symp. Proc. Vol. 1203 © 2010 Materials Research Society   1203-J14-05

Electron Emission from Diamond (111) p^+-i-n^+ Junction Diode

D. Takeuchi [1], T. Makino [1], H. Kato [1], M. Ogura [1], N. Tokuda [2], K. Oyama [1,3], T. Matusmoto [1,3], I. Hirabayashi [1], H. Okushi [1], S. Yamasaki [1,3]
[1]Energy Technology Research Center, AIST,
AIST-TC2-13, Umezono 1-1-1, Tsukuba, 305-8568, Japan
[2]Institute of Science and Engineering, Kanazawa University,
Kakuma-machi, Kanazawa 920-1192, Japan
[3]Graduate School of Pure and Applied Science, University of Tsukuba,
Tennodai 1-1-1, Tsukuba, 305-8577, Japan

ABSTRACT

We successfully observed electron emission from hydrogenated diamond (111) p^+-i-n^+ junction diodes. Here, p^+- and n^+-layers mean that the boron and phosphorous impurity concentrations in these layers are around 10^{20} cm^{-3}. The heavily doped layers play an important role to obtain high diode and emission currents. The emission started when the applying bias voltage was equal to the built-in potential, and the emission current reached to over 1 µA at room temperature operation. With taking into account our previous photoemission yield spectroscopy results and with the very high binding energy of free excitons of 80 meV in diamond, we suggested that the electron emission was derived from free excitons generated in the i-layer of the diodes.

INTRODUCTION

Electron emission from solid state devices is utilized in a variety of practical applications. Especially, electron emission devices with negative electron affinity (NEA) have been attracted a lot of researchers since high efficiency and unique properties of electron emission can be expected. We have investigated detailed photoelectron emission yield spectra from hydrogen-terminated (H-terminated) diamond surfaces with NEA, and concluded that useful electron emission from the bulk could not be obtained with a combination of n-type diamond and NEA. [1-3]

Recently Koizumi et al. successfully demonstrated electron emission from (111) p–n junctions with NEA. [4] The efficiency was recorded by more than 10% even though the device structure remained with enough room to develop. The device performances were mainly reported at higher temperatures of 200–300 °C to reduce resistance of n-type layer. The emission current of 1 µA at 200 °C was achieved with 100 V. He suggested that n-type layer behaved electron source, and the surface of p-type layer acted as NEA cathode. However, the electron emission mechanism, which gives ideas for optimization of the device structures, has not clearly been understood.

According to our previous works on photoelectron emission yield spectroscopy, we suggested that the electron emission from these diodes was derived from free excitons generated in the diodes, with taking into account the results of the photoemission yield experiments, [1,2] electron emission from a (001) diode with NEA, [5] and very high binding energy of free excitons of 80 meV. In addition, we aimed high-current at room temperature (RT) operation to develop diamond p-n diode based electron emitter with NEA surfaces for various applications.

To demonstrate our model, and to reach our goal with reducing series resistances, we attempted to fabricate diamond (111) p^+-i-n^+ junction diodes with p^+–layer on top, and investigate the I-V characteristics and electron emission properties at RT. [6] Here, p^+- and n^+-layers mean that the boron and phosphorous impurity concentrations in these layers are around 10^{20} cm^{-3}, where the Bohr radius of excitons becomes larger than a half of the average distance of each impurity atoms in diamond, and then the excitons can exist only in the i-layer of the diode. [7] The idea that such heavily doped layers can be used for p-n junctions has already been demonstrated by our group. [6,8] We thought that it is one of the most important properties of diamond to be able to use such heavily-doped layers for junctions.

EXPERIMENTAL DETAILS

For making p^+-i-n^+ junction diodes, each layer was grown by microwave-assisted plasma chemical vapor deposition (CVD). The substrate was a high pressure and high temperature synthesized Ib-type (111) diamond. The misorientation angle of this substrate was 2.7°. The growth condition of each layer is shown in Table. 1. The p^+-layer was grown by CVD system specified for heavily boron doping, where the incorporation of boron was controlled with C/H condition. [9] The i-layer was grown with 0.1% oxygen to reduce impurity incorporation. [10] The thickness and doping profiles of each p^+-, i-, and n^+-layer was measured by secondary ion mass spectroscopy (SIMS). Using photolithography and dry etching processes, mesa structures were prepared. Finally, Ti/Pt electrodes were deposited on both p^+- and n^+- layers.

For characterization, both the diode and the emission current-voltage characteristics were measured in ultra high vacuum (<10^{-9} Torr). These I-V characteristics were measured before and after hydrogenation by hydrogen radical irradiation with a hot filament system. The H-termination condition was established by means of total photoelectron emission yield spectroscopy with another diamond samples. [2] Before the measurement for the H-terminated case, the sample was annealed at 600 °C for 30 min to avoid the surface conductivity due to surface adsorbates. [11,12]

Table 1. Growth conditions of each layer of the p^+-i-n^+ junction diode.

Layer	C/H [%]	Impurity gas source	(Impurity gas) / CH$_4$ ratio [%]	Power [W]	Pressure [Torr]	Temperature [°C]	Duration [h]
p^+	0.2	Controlled by C/H. See text.		1200	50	930	50
i	0.5	O$_2$	0.1	3500	150	850	2
n^+	0.05	PH$_3$	0.025	750	75	800	12

RESULTS AND DISCUSSION

Sample structure

Figure 1 shows surface morphology of the sample after growth of the p^+-type top layer. The surface showed bunching steps without either non-epitaxial crystallite or pyramidal hillock. The bunching steps were observed just after growth of the n^+-type first layer. Thus, the p^+-i and i-n^+ interfaces must have this kind of macro-steps.

Figure 2 shows a result of SIMS for this sample. Each thickness of p^+-, i-, and n^+-layer was 0.26, 2.53, and 3.71 μm. The doping impurity concentrations of both p^+- and n^+-layers were almost the same as 10^{20} cm^{-3}, while the concentrations of boron, phosphorous, and nitrogen atoms in i-layer were under the detection limit of SIMS measurements. There was a step in the phosphorous profile in the n^+-layer in Fig. 2 due to independent two times deposition to obtain thick n^+-layer. The nitrogen atoms detected in the n^+-layer in Fig. 2, which seems to be due to the influence of hydrogen atoms (not shown) in this layer. There was an unintentionally induced peak of the profile of boron atom in i-n^+ interface.

Figure 3 (a) shows a top view of this sample after finishing all processes, and Fig. 3 (b)

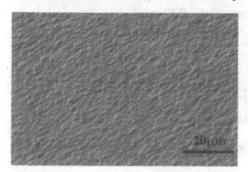

Fig. 1 Surface morphology of the sample after growth of the p^+-type top layer. The surface showed bunching steps without non-epitaxial crystallites.

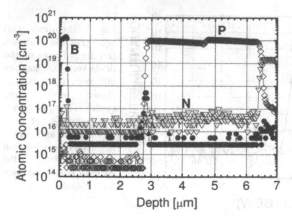

Fig. 2 A result of secondary ion mass spectroscopy (SIMS) for this sample. The doping impurity concentrations of both p^+ and n^+-layers were almost the same as 10^{20} cm^{-3}, while the concentrations of boron, phosphorous, and nitrogen atoms in i-layer were under the detection limit of SIMS measurement.

shows a schematic diagram of the cross section of a mesa structure in this sample. The mesa structure realizes a quasi-sandwich type p^+-i-n^+ junction diode.

The diode and the emission current-voltage characteristics

Fig. 4 shows the diode (Id) and the emission current (Ie) dependences on the applied voltage for the diode. Quantum efficiency (QE) is also shown with white circles. The electrode of p^+-layer was grounded, and then negatively biased region corresponded to the forward bias conditions for the diode. The diode showed very good rectification property. The rectification ratio around 20 V is 6-7 orders of magnitude. Before hydrogenation, no electron emission was observed even though the Id-V curve was almost the same as shown in Fig. 4. On the other hand, after hydrogenation, electron emission of 1-10 nA was observed around 5-10 V in forward bias conditions. It is important to notice that such electron emission starts around 4.5 V, which

(a)   (b)

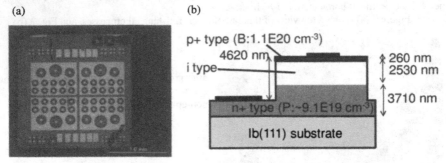

Fig. 3 (a) a top view of this sample after finishing all processes, (b) a schematic diagram of the cross section of this sample.

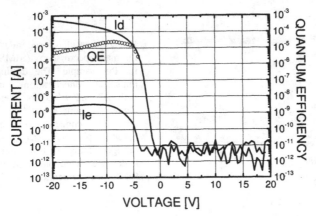

Fig. 4 The diode (Id) and emission currents (Ie) dependences on the diode voltage. Quantum efficiency (QE) is also shown with white circles. The electrode for p-type layer was grounded, and then negatively biased region corresponded to the forward bias conditions for the diode.

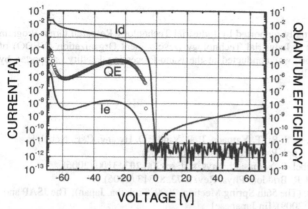

Fig. 5 The diode (*Id*) and emission currents (*Ie*) dependences of another diode in larger scale of the voltage applied to the diode. Quantum efficiency (*QE*) is also shown with white circles. The electrode for p-type layer was grounded, and then negatively biased region became forward bias conditions for the diode.

corresponds to the built-in potential of diamond *p-n* junction diode with the dopant impurities of boron and phosphorous. [6,8,10] Around these small forward bias conditions, p^+-i-n^+ junction region is in flat-band conditions, and then electrons and holes diffuse into *i*-layer without strong electric field. Under these conditions, electron-hole pairs form excitons. As denoted in introduction, heavily-doped layers could not contribute electron emission bia excitons. Thus, we thought that excitons-derived electron emission should be mainly involved in this result.

Fig. 5 shows the diode (*Id*) and emission currents (*Ie*) dependences of another diode in larger scale of the voltage applied to the diode. It showed more than 1 µA around 70 V with the quantum efficiency of 0.01% at RT operation. Such high net current of electron emission at room temperature operation was obtained from our structure. This fact indicates enough potential of free-exciton derived electron emission mechanism. Low efficiency and complex behavior of emission current observed in Fig. 5 seems to be due to the defects in junction region, which might be caused accumulation of boron atoms observed in Fig. 2.

CONCLUSIONS

We successfully observed electron emission from hydrogenated diamond (111) p^+-i-n^+ junction diodes. Electron emission started at around 4.5 V, which corresponded to the built-in potential of diamond *p-n* junction diode with the dopant impurities of boron and phosphorous. There the p^+-i-n^+ junction region was in flat-band conditions, and then electrons and holes formed excitons due to the very high binding energy of free excitons of 80 meV in diamond. In addition, heavily-doped layers could not contribute to electron emission bia excitons. Thus, we thought that excitons-derived electron emission should be mainly involved in this result. The diode showed more than 1 µA electron emission current around 70 V with the quantum efficiency of 0.01% at room temperature operation. Such high net current of electron emission was obtained from our structure, which indicates enough potential of free-exciton derived electron emission mechanism.

ACKNOWLEDGMENTS

This research was partially supported by Industrial Technology Research Grant Program in 2008 from New Energy and Industrial Technology Development Organization (NEDO) of Japan, and a part of this work was conducted at the Nano-Processing Facility, supported by IBEC Innovation Platform, AIST.

REFERENCES

1. D. Takeuchi, H. Kato, G. S. Ri, T. Yamada, P. R. Vinod, D. Hwang, C. E. Nebel, H. Okushi, S. Yamasaki, Appl. Phys. Lett. **86**, 152103 (2005).
2. D. Takeuchi, C. E. Nebel, S. Yamasaki, phys. stat. sol. (a) **203**, 3100 (2006).
3. S. J. Sque, R. Jones, P. R. Briddon, Phys. Rev. B **73**, 85313 (2006).
4. S. Koizumi, Ext. Abst. (The 56th Spring Meeting, 2009, Tsukuba, Japan); The JSAP and Relat. Soc. No.2, 612 (2009). [in Japanese]
5. D. Takeuchi, T. Makino, S.-G. Ri, N. Tokuda, H. Kato, M. Ogura, H. Okushi, S. Yamasaki, Appl. Phys. Express. **1**, 15004, (2008).
6. H. Oyama, S.-G. Ri, H. Kato, M. Ogura, T. Makino, D. Takeuchi, N. Tokuda, H. Okushi, S. Yamasaki, Appl. Phys. Lett. **94**, 152109 (2009).
7. D. Takeuchi, M. Ogura, N. Tokuda, H. Okushi, S. Yamasaki, phys. stat. sol. (a) **206**, 1991 (2009).
8. T. Makino, H. Kato, S.-G. Ri, S. Yamasaki, H. Okushi, Diamond Relat. Mater. **17**, 782 (2008).
9. D. Takeuchi, S. Yamanaka, H. Watanabe, H. Okushi, phys. stat. sol. (a) **186**, 269 (2001).
10. T. Makino, N. Tokuda, H. Kato, M. Ogura, H. Watanabe, S.-G. Ri, S. Yamasaki, H. Okushi, Jpn. J. Appl. Phys. **45**, L1042 (2006).
11. D. Takeuchi, M. Riedel, J. Ristein, L. Ley, Phys. Rev. B, **68**, 041304(R) (2003).
12. M. Riedel, J. Ristein, L. Ley, Phys. Rev. B, **69**, 125338 (2004).

Device Applications

(5) Electronics Applications

Mater. Res. Soc. Symp. Proc. Vol. 1203 © 2010 Materials Research Society   1203-J15-04

RF Power Performance Evaluation of Surface Channel Diamond MESFETs

M. C. Rossi[1], P. Calvani[1], G. Conte[1], V. Camarchia[2], F. Cappelluti[2], G. Ghione[2], B. Pasciuto[3], E. Limiti[3], D. Dominijanni[4] and E. Giovine[4]

[1] Department of Electronic Engineering, Università di Roma Tre, Roma, Italy
[2] Department of Electronics, Politecnico di Torino, Torino, Italy
[3] Department of Electronic Engineering, Università di Tor Vergata, Roma, Italy
[4] IFN-CNR, Roma, Italy

ABSTRACT

Large-signal radiofrequency performances of surface channel diamond MESFET fabricated on hydrogenated polycrystalline diamond are investigated. The adopted device structure is a typical coplanar two-finger gate layout, characterized in DC by an accumulation-like behavior with threshold voltage $V_t \sim$ 0-0.5 V and maximum DC drain current of 120 mA/mm. The best radiofrequency performances (in terms of f_T and f_{max}) were obtained close to the threshold voltage. Realized devices are analyzed in standard class A operation, at an operating frequency of 2 GHz. The MESFET devices show a linear power gain of 8 dB and approximately 0.2 W/mm RF output power with 22% power added efficiency. An output power density of about 0.8 W/mm can be then extrapolated at 1 GHz, showing the potential of surface channel MESFET technology on polycrystalline diamond for microwave power devices.

INTRODUCTION

Diamond is, in principle, the highest performance widegap semiconductor, with outstanding electronic and thermal properties (high breakdown field, large carriers mobility and saturated velocity, thermal conductivity about 15 times larger than GaN and 50 than GaAs). This makes it an attractive material for high power radiofrequency (RF) and microwave electron devices, and a good candidate as a vacuum tube replacement in high power, high frequency applications like communications satellite and radar power stages. Although the properties of synthetic diamond have been known from many years (with the first diamond-based FETs demonstrated at the beginning of 90s) only recently significant technology advances in growth techniques for single-crystal and polycrystalline diamond have fostered the research on high-performance diamond electronics. Of particular significance is, at the present stage, the development of active devices on large area (4 inch wafers) polycrystalline diamond films, with 100-200 μm grain size, grown on different substrates, while single-crystal diamond substrates films are limited to smaller areas (a few square millimeters). Available approaches for the control of diamond conductivity rely mainly (up to now) on p-type doping, either through extrinsic doping with boron or exploiting hydrogen (H) surface termination, which induces a quasi-2D hole channel a few nanometers below the surface. Both approaches are being pursued in order to realize high speed power FETs [1], although the development of delta-channel FETs (based on boron delta-doped channels) is today well behind the one of H-terminated surface-channel MESFETs which have demonstrated cut-off frequencies in the GHz range. In particular,

H-terminated FETs with 0.1 μm gate length on polycrystalline diamond reaching 45 GHz cut-off frequency have been reported [2]. Such good performances are however limited by stability problems affecting surface channel FET operation. In order to overcome this problem, suitable surface passivation processes able to prevent channel conductivity degradation, have been recently suggested [3] as well as treatments inducing time-stable transfer doping are currently under investigation. Such a perspective justifies the interest toward H-terminated FETs, typically showing promising small-signal performances. At variance, only few examples of RF power measurements have been reported so far in the literature, limited to comparatively low frequency (1 GHz), and only for single-crystal diamond FETs, achieving a record performance of 2 W/mm [4]. In this paper, we present large-signal RF measurements of submicron H-terminated FETs grown on polycrystalline diamond up to 2 GHz, showing the potential of such a substrate for the development of diamond-based microwave power devices.

DEVICE STRUCTURE AND REALIZATION

MESFET devices were fabricated on 250 μm thick diamond substrate of 1.0×1.0 cm^2, with surface roughness less than 50 nm and showing an average grain size of about 80-100 μm, supplied by Element Six. The smoothed film surface was treated in a high temperature microwave hydrogen plasma by using a Astex PDS19 CVD system (5 kW power, 2.45 GHz frequency) in order to induce a p-type surface conductivity to be used for the realization of the transistor channel. As shown in fig. 1 (left), a typical butterfly shaped layout with an optimized coplanar two fingers gate RF layout, able to reduce the input-output impedance mismatch, has been designed in order to assure high frequency device operation. The device active area was defined by reactive ion etching (RIE) in oxygen and argon plasma, achieving electrical isolation among FET structures. It is worth to note that the isolation pattern was established in order to force the drain to source current to flow underneath the gate, thereby avoiding parasitic resistance effects. Deep sub-micron I-shaped gate electrode with length L$_G$ down to 200 nm and widths in the range 50 - 200 μm have been realized by a single-layer electron beam lithography. Gold and aluminum have been used for the realization of the drain and source ohmic contacts, and the gate Schottky contact, respectively.

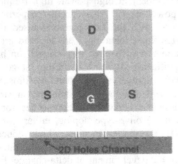
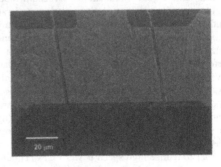

Fig. 1 Left: Schematic of the device structure Right: Scanning electron microscope (SEM) image of the device active area.

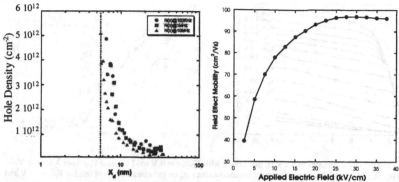

Fig. 2 Left: Surface hole density as evaluated from C-V measurements. Right: Channel mobility *vs.* electric field as estimated from typical devices trans-characteristic.

A self-aligned gate technology was employed in order to reduce the drain to source effective channel length down to less than 2.5 μm. A detail of the realized devices on polycrystalline diamond is reported in fig. 1 (right). The channel, located about 6 nm under the surface, exhibited complete activation already at room temperature and a surface carrier concentration around 10^{13} cm^{-2}, as evaluated from C-V measurements and reported in fig. 2 (left). Channel mobility, estimated from measured trans-characteristic (see fig. 2 (right)) and confirmed by Hall-mobility measurements, was about 95 cm^{2}V^{-1}s^{-1}.

RESULTS AND DISCUSSION

DC and RF small-signal characteristics

Devices (two 0.2 μm fingers with total gate periphery of 100 μm) were first characterized in DC conditions, showing a maximum drain current of about 120 mA/mm (see fig. 3 (left)). The trans-characteristics and gate bias dependence of the extrinsic transconductance at V_{DS} = -8 V and V_{DS}=-14 V are reported in fig. 3 (right). The enhancement-like behavior is characterized by a threshold voltage around 0.5 V and maximum extrinsic transconductance around 4 0mS/mm. The small-signal RF characterization was carried out in the range 0.1 - 20 GHz at different quiescent points. Two examples of measured input and output scattering parameters S_{11} and S_{22} are reported in Fig. 4 at two different bias points: V_{DS}=-14 V, V_{GS}=-0.9 V close to the maximum g_m (Fig. 4 (left)) and V_{DS}=-14 V, V_{GS}=0.3V (Fig. 4 (right)) close to the threshold. Multi-bias small-signal characterizations have been exploited in order to develop a large-signal dynamic equivalent circuit of the device, modeling the bias dependence of the reactive elements. Fig. 5 (left) shows the extracted behavior of the intrinsic transconductance and gate-to-source capacitance as a function of V_{GS}, at V_{DS}=-14 V. The bias-dependence of the small-signal intrinsic g_m is close to the one obtained from DC characterization, indicating that these devices are not affected by relevant dispersive effects. The difference between the DC and small-signal behavior may be ascribed to the parasitic elements. The gate-to-source capacitance increases as the device (i.e. the 2D hole channel) is turned on, as expected for MOSFETs or heterostructure-based FETs.

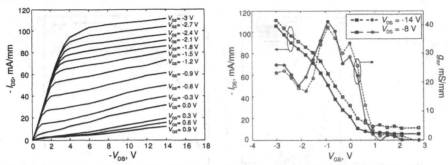

Fig. 3 Left: Output characteristics of a 2 x 50 μm device with V_{DS} from 0 V to -14 V and V_{GS} from 3 V to -3 V (upper curve). Right: Trans-characteristics and transconductance g_m on the same device of left for V_{DS} = - 8 V and V_{DS} = - 14 V.

Fig. 5 (right) reports the extracted behavior of C_{GS} as a function of V_{DS} at V_{GS}=-0.9 V. On the other hand, the intrinsic transconductance turned out to be mildly dependent on V_{DS}. The extracted bias dependence of g_m and C_{GS} suggests a dependence of the cut-off frequency (f_T) and maximum oscillation frequency (f_{max}) on V_{GS}, V_{DS} in qualitative agreement with the measured frequency response shown in Fig. 3. In fact, as already reported in literature (see e.g. [5]), these devices show optimum f_T and f_{max} close to the threshold and a slight dependence on V_{DS}. Slightly higher cutoff frequency was obtained with the same process, by exploiting higher quality (and more expensive) polycrystalline substrates, see [6].

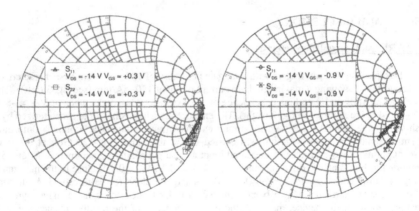

Fig. 4 Input (S_{11}) and output (S_{22}) reflection coefficients measured at V_{DS}=-14 V, V_{GS} =0.3 V (left) and V_{DS} =-14V, V_{GS} =-0.9 V (right).

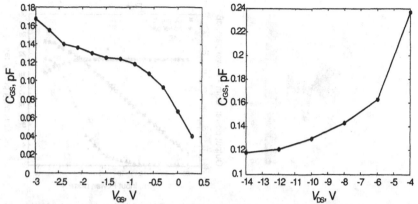

Fig. 5 Left: Gate-to-source capacitance vs. V_{GS} at $V_{DS} = -14$ V. Right: gate-to-source capacitance vs. V_{DS} at $V_{GS} = -0.9$ V (right), as extracted from multi-bias S-parameters.

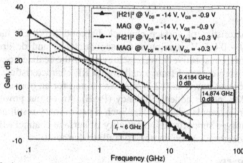

Fig. 6 Current gain and Maximum Available Gain (MAG) for $V_{DS} = -14$V, $V_{GS} = -0.9$ V (solid line) and $V_{DS} = -14$V, $V_{GS} = +0.3$ V (dashed line).

RF power performance

The evaluation of RF power performance was carried out at 2 GHz, under standard class A operation ($V_{DS} = -14$ V, $V_{GS} = -0.9$ V), employing a nonlinear test bench based on the active load pull technique. An example of load pull map is shown in fig. 7 (left). From this measurement, we found an optimal load for maximum PAE of $\Gamma_L = 0.79$, $4°$. Fig. 7 (right) shows the results of a power sweep with the device terminated on the optimum load. The maximum output power density is 0.2 W/mm with 22% power added efficiency (PAE), and linear power gain of 8 dB. The extrapolated output power density at 1 GHz is about 0.8 W/mm; such value, though still much lower than the expected material limit, points out the potential of this technology on polycrystalline diamond if compared e.g. to the record performance on single-crystal diamond [2] quoted above.

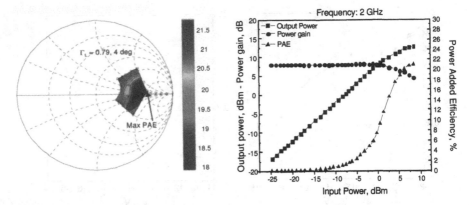

Fig. 7 Left: Load-pull map of the 2x50 μm device operating at 2 GHz, and biased in class A (V_{GS} =-0.9 V, V_{DS} =-14 V). The marker shows the location of the optimum load for maximizing the PAE. Right: Corresponding power sweep on the PAE optimum load (Γ_L=0.79, 4°). Power Gain (dots), output power (squares) and PAE (triangles).

CONCLUSIONS

The RF characteristics of surface channel diamond MESFET have been investigated. Threshold voltage $V_t \sim$ 0-0.5 V, maximum transconductance and dc drain current around 40 mS/mm and 120 mA/mm, respectively, were achieved for the realized devices. Optimum f_T = 6 and f_{max} = 14.8 GHz were detected from small signal RF characteristics. From the large signal point of view, an RF output power of 0.2 W/mm was found for an operating frequency of 2 GHz with 8 dB gain and 22% power added efficiency (PAE). An output power density of about 0.8 W/mm have been then extrapolated at 1 GHz. Such a preliminary result achieved for a not fully optimized device prototype supports the potentiality of surface channel MESFET technology for microwave power devices realization on polycrystalline diamond.

ACKNOWLEDGMENTS

Support of the CNT4SIC project of Regione Piemonte is acknowledged.

REFERENCES

1. E. Kohn, A. Denisenko, *Thin Solid Films* **515**, 4333-4339 (2007).
2. K. Ueda, M. Kasu, Y. Yamauchi, T. Makimoto, M. Schwitters, D. J. Twitchen, G. A. Scarsbrook, S. E. Coe, *IEEE Electron Dev. Lett.* **27**, 7,570-572 (2006).
3. D. Kueck , S. Jooss, E. Kohn, *Diam. Relat. Mater.***18**, 1306-1309 (2009).
4. M. Kasu, K. Ueda, H. Ye, Y. Yamauchi, S. Sasaki, T. Makimoto, *Electron. Lett.* **41**, 22 (2005).
5. M. Kubovic, M. Kasu, I. Kallfass, M. Neuburger, A. Aleksov, G. Koley, M.G. Spencer, E. Koh, *Diam. Relat. Mater.* **13**, 802-807 (2004).
6. P. Calvani, A. Corsaro, M. Girolami, F. Sinisi, D. M. Trucchi, M. C. Rossi, G. Conte, S. Carta, E. Giovine, S. Lavagna, E. Limiti, V. Ralchenko, *Diam. Relat. Mater.* **18**, 786-788 (2009).

Mater. Res. Soc. Symp. Proc. Vol. 1203 © 2010 Materials Research Society 1203-J04-01

Study and Optimization of Silicon-CVD Diamond Interface for SOD Applications

Jean-Paul Mazellier[1,3], Jean-Charles Arnault[2], Mathieu Lions[2], François Andrieu[1], Robert Truche[1], Bernard Previtali[1], Samuel Saada[2], Philippe Bergonzo[2], Simon Deleonibus[1], Sorin Cristoloveanu[3], Olivier Faynot[1]

[1] CEA, LETI, MINATEC, Innovative Devices Laboratory, F-38054 Grenoble, France
[2] CEA, -LIST, Diamond Sensors Laboratory, F-91191 Gif-sur-Yvette, France
[3] IMEP-LAHC MINATEC, F38016 Grenoble, France

ABSTRACT

With respect to Silicon-on-Diamond approaches as an alternative to SOI where diamond is used as the buried dielectric, we have in recent works demonstrated the feasibility of a novel approaches where the CVD diamond layer is grown on silicon using Bias Enhanced Nucleation (BEN) over large area substrates, then smoothed and assembled to successfully enable the fabrication of first prototypes of silicon-on-diamond substrates. The key novelty to those SOD substrates were that only a very thin box dielectric diamond layer is used (typically from 150 to 500nm thick), as required by the current SOI technology. However we had also observed that the silicon-diamond interface quality to be sensitive to the nature of the nucleation interface. Thus the current contribution here studies the chemical nature of various capping materials used to solve the issue of electrical defects in case of direct silicon-diamond interface and at the same time to enable the whole system to benefit from the high thermal conductivity of diamond when compared to other standard electrical insulating materials.

INTRODUCTION

Nowadays, the semiconductor industry is facing an exciting challenge concerning its future. Not only transistor size shrinking is to be overcome but also innovative integration of new architectures and functionalities are to be managed. This advanced integration requires solutions to problems that were not addressed up to now. Thermal management at the device scale is part of this new era that needs smart solutions. Especially concerning Silicon-On-Insulator (SOI) substrates: this technology offers lots of advantages for advanced devices such as easy processing but also enhanced static and dynamic electric control. But the standard Buried OXide (BOX) layer is a supplementary issue in terms of thermal management ($k_{SiO2} = 1.4$W/mK when $k_{Si} = 150$W/mK). Some works have already focused on alternative material for BOX replacement, by integrating CVD diamond layers, but without any proof of amelioration [1]. We have in an earlier communication already demonstrated the feasibility of the replacement of the standard BOX in an SOI substrate with a thin CVD diamond layer, thus forming Silicon-On-Diamond (SOD) substrates [2,3]. The integration scheme used implies the growth of diamond directly on the silicon that will represent the top active layer of the final SOD substrate. However, when Bias enhanced Nucleation is used, it is likely that the ion bombardment at the first initial stage of growth result in an increased localized space charge at this interface that may ultimately alter the SOI substrate performances. To get further insights into this mechanism, we focus in the current study on the nature of the interface between silicon and diamond. Two ways have been investigated and compared: direct growth of diamond on silicon or use of an intermediate capping between both materials [4].

EXPERIMENT

We have grown diamond on three different types of wafers. Description of these wafers is given in Table I.

Table I. Details about the structure of each studied sample.

Sample	#A	#B	#C
Substrate	Silicon (100), Arsenic doped (1-10 Ω.cm)		
Structure	Si (no capping)	Si/SiO₂ 10nm/ polysilicon 10nm	Si/SiO₂ 10nm/Si₃N₄ 5nm

Each of them is composed of a (100) silicon wafer (Arsenic doped, resistivity 1-10 Ω.cm). Sample #A consists of a sole silicon wafer, without capping. For sample #B and #C, a thermal oxidation (97% O_2 + 3% HCl, 1050°C) was performed to form a 10nm layer of SiO_2. Then for sample #B, an other 10nm of polysilicon was deposited using LPCVD at 550°C. Also, for sample #C, a 5nm of Si_3N_4 was deposited by LPCVD at 780°C.

For each samples (see Table I), submicron diamond layers were grown using Chemical Vapour Deposition assisted by Microwave Plasma (MPCVD) at 2.45 GHz. The first step consists of exposing the wafer to a CH_4/H_2 plasma in order to stabilize the temperature and to form a thin silicon carbide layer. The injected microwave power is adjusted at 950W, the pressure at 26 mBar, the gas mixture at 10vol % CH_4 and 90 vol% H_2 and the gas flow is 120 sccm. This step lasts 5 min and the temperature reaches 1013 K. The second step is the diamond nucleation stage performed using BEN technique [5,6]. For the BEN step, the process parameters are the same than the previous step except that a voltage of -307 V is applied to the substrate during 15 min to initiate nucleation at a density above 10^{11} sites/cm^{-2} [7].

After this nucleation BEN stage, the gas injection conditions and microwave power were modified to become [99,4 Vol.% H_2 / 0,6Vol. % CH_4] with a flow of 250sccm and a pressure of 35mBar, the input microwave power being adjusted at 1100W. In these conditions the substrate temperature is 1123 K and the diamond growth rate reaches 220nm/h.

The surface chemistry has been investigated using in-situ X-ray Photoelectron Spectroscopy (XPS). XPS was performed using an OMicron XPS spectrometer equipped with an Al Kα monochromatized anode (hν = 1486.6 eV). The binding energy scale was calibrated versus the Au 4f 7/2 peak located at 84.0 eV [8].

For electrical characterization, we have annealed the diamond layers at 400°C during 1h in ambient atmosphere in order to deshydrogenate the grown surface. Then a 200nm layer of gold has been evaporated through a molybdenum grid (square shaped, 450μm side) for electrical contact fabrication. The silicon substrate is used as back electrode. Current-Voltage and Capacitance-Voltage characteristics have been measured using an Agilent 4156C Semiconductor Analyzer and HP-4282 LCR meter respectively.

For the thermal characterization, we used the 3ω method [9]. For this purpose, we have formed a 4 points metallic probe by evaporating 100nm of aluminum on the diamond surface. Then the measurement structures were formed using photolithography, IBE etching and oxygen based plasma stripping of the resist. Note that before aluminum deposition, diamond layers were planarized using plasma based technique [10]. By smoothing the surface of the diamond, we ensure the thermal measurements to be accurate.

DISCUSSION

Diamond layers and substrate physical characterization

After submicron (about 400nm) diamond layer formation on each samples, we have observed the morphology of the different films using Scanning Electron Microscopy (SEM). We report the comparison of cross sections for the three samples on Figure 1. We see that all films exhibit the same columnar structure with comparable grain sizes. One important thing is that the capping layers in cases #B and #C remain unaffected after the diamond process. In particular, the SiO_2 layer integrity seems totally preserved even if the diamond process uses a highly hydrogen concentrated plasma, a reducing environment that etches SiO_2 efficiently.

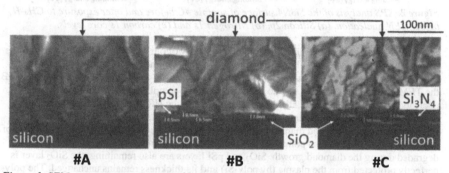

Figure 1. SEM cross section of diamond layers grown on different samples. We have focused on the interface between the silicon substrate and the diamond film. All samples are represented at the same scale.

In the case of sample #A, the standalone silicon wafer, in situ XPS analyses in series with diamond nucleation had already been reported [11] and results had demonstrated the etching of the native SiO_2 layer with consecutive formation of a non stœchiometric silicon carbide layer of about 3nm thick. We assume this mechanism to take place also for sample #B where the top layer is polysilicon, thus nearly here almost the same material as the monocrystalline silicon of sample #A. On the contrary, in the case of sample #C, Si_3N_4 is the top material and its interaction with plasma was not previously investigated. XPS analysis of this sample before and after H_2/CH_4 plasma exposure during the stabilization step was performed in situ and our results are summarized on Figure 2. By focusing on Si2p and N1s core levels, we can observe that the plasma exposure show almost no effect on the Si_3N_4 overlayer, demonstrating no carbide formation of the top film on contrary to case #A). At the C1s core level, we can observe the deposition of a carbon layer on sample #C. The atomic concentration for carbon is measured at 10.7 at. %.

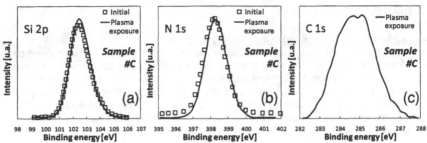

Figure 2. XPS analysis of the Si3N4 surface of sample #C before and after exposure to CH₄-H₂ before BEN nucleation. (a) Silicon 2p, (b) Nitrogen 1s and (c) carbon 1s core levels.

In order to get a better insight into the structural integrity of our samples, and particularly of the SiO₂ capping and underlying silicon substrate, we have performed TEM and HRTEM cross section observations on each sample. The results are presented on Figure 3. For cases #A and #B, the TEM observation confirms that no degradation of the sample occurred during diamond nucleation and growth. On the silicon wafer, sample #A, the HRTEM view demonstrates the absence of crystalline defects at the substrate interface (Figure 3.A down). We can notice that interface between silicon and diamond presents a roughness on the order of 1nm. For the case of sample #B, the HRTEM also points out that the underlying monocrystalline silicon is not degraded during the diamond growth. SiO₂ and pSi layers are also remaining: the SiO₂ layer is perfectly protected from the plasma (by poly-Si) and its thickness remains unchanged. The poly-Si is partially etched during the process (almost due to the carbide layer formation as in the case of sample #A) but at least a 5nm of poly-Si remains after diamond growth. For sample #C, the TEM observation reveals clear traces of defects starting from the first nanometers of the diamond films and ending within the silicon substrate, crossing the whole SiO₂/Si₃N₄ capping. These traces are quite dense and uniformly distributed over the whole sample. HRTEM view demonstrates that these traces end in the silicon substrate, where this last one has been amorphized. The localization of these traces let us think that these degradations occurred during the BEN nucleation: during biasing of the sample and charges exchange, the SiO₂/Si₃N₄ capping may disturb the charge spreading, involving local high field formation and thus electrical breakdown, allowing the local silicon amorphization. This behavior has also been observed on SOI substrates where buried SiO₂ (145nm thick) was acting as electrical insulator.

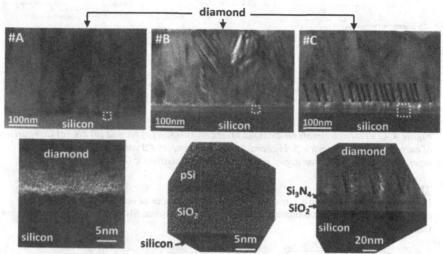

Figure 3. TEM (left) and HRTEM (right) cross section of different diamond layers grown on the three studied samples.

Electrical characterization

Capacitances and I(V) measurements were performed on all three samples. For sample #C, non reproducible characteristics (not shown here) were obtained. We relate this to the structural degradation of the sample as demonstrated previously by TEM and HRTEM observations. We kept focus of the study on samples #A and #B. C(V) measurements were conducted using voltage sweeping from +40V to -40V to +40V (silicon substrate kept at ground voltage):. The oscillator frequency was kept constant at 1MHz. Our data are presented on Figure 4.a and 4.b. In the case of sample #A, an hysteresis is clearly evidenced in the accumulation region which demonstrates the presence of electrically active defects at the silicon-diamond interface. On the other hand, clearly for sample #B, no hysteresis could be observed. In fact diamond-silicon interface traps have already been pointed out in other studies where nanoseeding was employed [4,12], contrarily to our BEN technique. In the same way, one study pointed out the interest of capping silicon to maintain the electrical quality of Si-SiO$_2$ interface, but also considering nanoseeding technique. For both cases, desertion can be obtained in the silicon film and the 1/C^2 vs. voltage plot (inset of Figures 4.a and 4.b) demonstrates linear behavior enable the extraction of the doping concentrations of 4.9×10^{15}/cm^3 and 2.5×10^{15}/cm^3 respectively, in perfect agreement with resistivity specifications. Note that we have extracted a relative dielectric permittivity of 5.2±0.2 of diamond from the accumulation plateau, a value very close to that of bulk diamond (5.4 to 5.7). Comparing current density vs. electric field (Figure 4.c) of sample #B with a diamond layer grown on highly doped silicon substrate, we can see very similar behaviors, demonstrating that electrical parameters of diamond are not influenced by the presence of the capping.

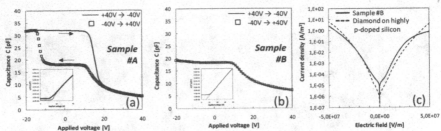

Figure 4. C(V) measurements on diamond layers on samples (a) #A and (b) #B. (c) Comparison
of current density vs. electric field characteristics of samples #B with a submicron diamond
layer grown on highly boron doped silicon substrate (resistivity < 0.01 Ω.cm)

Thermal measurements

Using the 3ω method [9], we extracted the thermal resistances of our different samples and
compared them with other insulating materials such as SiO_2 and Si_3N_4 of comparable thickness
(about 400nm).

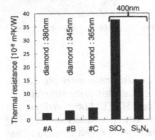

Figure 5. Comparison of thermal resistances of our
different samples and other insulating materials. In each
case, the total thickness of the stack is given.

Comparing samples #A, #B and #C we observe that the introduction of the extra capping indeed
induces an increase of the thermal resistance, that can be attributed to the different thermal
interfaces at each materials junction. Further, the polysilicon incorporating capping (#B) is more
thermally efficient than the Si_3N_4 one (#C). However, this increase in thermal resistance remains
very low when compared with other insulating materials such as SiO_2 or Si_3N_4 for all our
samples (#A, #B and #C): they exhibit at least a factor of 3 in thermal resistance reduction when
compared to the 400nm of Si_3N_4. This ratio reaches a factor of 10 when compared to a 400nm
thick SiO_2 layer.

CONCLUSIONS

We have investigated the growth of CVD diamond layer by MPCVD coupled with BEN
technique. We have demonstrated that this process can lead to diamond growth on SiO_2/Si_3N_4
capped silicon substrate but one detrimental effect is that the BEN step induces a significant
electrical breakdown in the insulating stack that degrades the silicon. On the other hand, using an
SiO_2/polysilicon capping allows to suppress this electrical failures and demonstrate perfect
integrity of both underlying silicon and SiO_2. From the electrical point of view, the integration of
the SiO_2 layer allows to recover the electrical quality as good as that of the Si-SiO_2 one, when

compared to a direct Si-diamond interface that shows high defect density. Eventually, the integration of a capping layer before the diamond growth, even if it shows slightly higher resistance, does not degrade significantly the thermal resistance of the whole system and the benefit of diamond is still clearly observable compared to other thin film electrical insulators.

ACKNOWLEDGMENTS

The authors acknowledge the French Carnot Institute for fundings of this work.

REFERENCES

[1]   B. Edholm, L. Vestling, M. Bergh, S. Tiensuu and A. Söderbärg, "Silicon-On-Diamond MOS-Transistors with Thermally Grown Gate Oxide" Proceedings IEEE International SOI Conference (1997)
[2]   J. Widiez, M. Rabarot, S. Saada, J.-P. Mazellier, J. Dechamp, V. Delaye, J.-C. Roussin, F. Andrieu, O. Faynot, S. Deleonibus, P. Bergonzo, L. Clavelier, "Fabrication of Silicon-On-Diamond (SOD) Substrates by either Bonded and Etched Back SOI (BESOI) or the Smart-Cut Technology". Solid-State Electronics *to be published* (2010)
[3]   J.-P. Mazellier, J. Widiez, F. Andrieu, M. Lions, S. Saada, M. Hasegawa, K. Tsugawa, L. Brevard, J. Dechamp, M. Rabarot, V. Delaye, S. Cristoloveanu, S. Deleonibus, P. Bergonzo, O. Faynot, "Diamond integration for thermal management in thin Silicon MOSFETs". Proceedings IEEE International SOI Conference (2009)
[4]   B. Edholm, J. Olsson, N. Keskitalo and L. Vestling, "Electrical investigation of the Silicon/Diamond Interface". Microelectronic Engineering **36**, p. 245-248 (1997)
[5]   S. Yugo, T. Kanai, T. Kimura and T. Muto, "Generation of diamond nuclei by electric field in plasma chemical vapor deposition". Applied Physics Letters **58**, p. 1036-1038(1991)
[6]   S. Saada, J.-C. Arnault, L. Rocha and P. Bergonzo, "Synthesis of Sub- Micron Diamond Films on Si(100) for Thermal Applications by BEN-MPCVD". Material Research Society Symposium **956** (2007)
[7]   M. Lions, S. Saada, J.-P. Mazellier, F. Andrieu, O. Faynot, P. Bergonzo, "Ultra-thin nanocrystalline diamond films (<100 nm) with high electrical resistivity". Physica Status Solidi (RRL) **3**, p. 205-207 (2009)
[8]   M.P. Seah, "Post-1989 calibration energies for X-ray photoelectron spectrometers and the 1990 Josephson constant ". Surface and Interface Analysis **14**, p. 488 (1989).
[9]   S.-M. Lee and D. G. Cahill, "Heat transport in thin dielectric films". Journal of Applied Physics **81**, p. 2590-2595 (1997)
[10]   M. Rabarot, J. Widiez, S. Saada, J.-P. Mazellier, J.-C. Roussin, J. Dechamp, P. Bergonzo, F. Andrieu, O. Faynot, S. Deleonibus, and L. Clavelier, "Silicon-on-Diamond layer integration by wafer bonding technology". Diamond and Related Materials *to be published* (2010)
[11]   S. Saada, J.C. Arnault, N. Tranchant, M. Bonnauron, P. Bergonzo, "Study of the CVD process sequences for an improved control of the Bias Enhanced Nucleation step on silicon". Physica Status Solidi (a) **204**, p. 2854-2859 (2007).
[12]   A. Jauhiainen, S. Bengtsson and O. Engström, "Steady-state and transient current transport in undoped polycrystalline diamond films". Journal of Applied Physics **82**, p. 4966-4976 (1997).

Device Applications

(6) Optical Applications

Mater. Res. Soc. Symp. Proc. Vol. 1203 © 2010 Materials Research Society   1203-J13-01

The Outlook for Diamond in Raman Laser Applications

Richard P. Mildren
MQ Photonics Research Centre
Department of Physics
Macquarie University, New South Wales, 2109, Australia

ABSTRACT

Efficient and practical Raman lasers based on single crystal diamond are now realizable owing to the availability of optical quality crystals grown by chemical vapour deposition (CVD). In this paper, the performance characteristics of CVD-diamond Raman lasers are summarized and the results compared to those for more established Raman materials. The outlook for diamond Raman lasers is discussed and key challenges for material development highlighted.

INTRODUCTION

Much like the electronic industry, lasers are being developed with ever increasing power, speed and frequency range. Almost all fields of science and technology now benefit from laser technology in some way and demand a range of specifications that will include output wavelength, beam power, temporal format, coherence and system parameters such as footprint and efficiency. Thus there is an ongoing search for alternatives to the optical gain material that is fundamental to laser performance. Diamond is highly attractive as a laser material as it promises capabilities well beyond that possible from other materials in accordance with its extreme properties. Most diamond laser research to date has concentrated on doped diamond for color center lasers [1], semiconductor diode lasers [2] and rare earth doped lasers [3]. Success has been very limited except from perhaps color center lasers relying on the nitrogen vacancy that have been demonstrated with an optical-to-optical conversion efficiency of 13.5% [1]. The major challenge for diamond as a laser host is the incorporation of suitable concentrations of color centers or active laser ions into the tightly bonded lattice either by substitution or interstitially. On the other hand, Raman lasers rely on stimulated scattering from fundamental lattice vibrations and thus do not require doping [4,5]. Though the principle of optical amplification is distinct from conventional lasers that rely on a population inversion, in many ways Raman lasers have similar basic properties to other laser-pumped lasers. Raman lasers can be thought functionally as laser converters that bring about a frequency downshift and improved beam quality. Their development has been most often driven by the need for laser wavelengths that are not fulfilled by conventional laser media and find use in a diverse range of fields such as in telecommunications, medicine, bio-diagnostics, defence and remote sensing.

Synthetic (CVD) single crystal diamond has become available in the last few years with size, optical quality and reproducibility well suited for implementation in Raman lasers. Diamond's starkly different optical and thermal properties compared to "conventional" materials are of substantial interest for extending Raman laser capabilities. Diamond has the highest Raman gain coefficient of all known materials (approximately 1.5 times higher than barium nitrate) and outstanding thermal conductivity (more than two orders of magnitude higher than most other Raman crystals) and optical transmission range (from 0.230 μm and extending to beyond 100 µm, with the exception of 3-6 µm). These properties herald promise for substantially raising average output power of a small device and extending the spectral reach beyond the visible and near infrared.

There is a considerable knowledge base in crystalline Raman lasers developed since the early 1990s that can be applied to synthetic diamond. In this paper we provide some basic background on the design and performance characteristics of common crystal Raman laser architectures including in particular the external cavity Raman laser configuration. We then compare the performance of diamond external cavity Raman lasers to that seen in other materials and foreshadow some of the interesting opportunities for diamond Raman laser development.

Background on Crystalline Raman Lasers

Raman lasers rely on the phenomenon of stimulated Raman scattering (SRS) for optical amplification in the laser resonator. Input pump photons of frequency ω_p excite a normal mode of vibration in the crystal lattice and the remaining energy is carried away as Stokes shifted photons of frequency ω_s. Note that stimulated anti-Stokes Raman scattering to generate up-shifted photons is typically an inefficient process that requires a more sophisticated formalism not covered in this paper. The probability for Raman scattering is higher for materials that change in polarizability α with small displacements dq in the lattice vibration i.e., for large $d\alpha/dq$. $d\alpha/dq$ is a measure of the amount of distortion experienced by the electron cloud as a result of the incident light and its square is directly proportional to the spontaneous Raman cross-section. SRS requires the interaction of a Stokes photon with two pump photons and is thus a third-order nonlinear optical process (like third harmonic generation, four-wave mixing and two-photon absorption). Amplification of the Stokes field intensity I_s as it propagates through the Raman medium on the z-axis is given by

$$dI_s/dz = g.I_p.I_s \tag{1}$$

where the gain coefficient g is proportional to $(d\alpha/dq)^2$, the Stokes frequency ω_s and under steady-state conditions[†] the time for coherent lattice phonons to dissipate in the material (T_2):

$$g = k/m.\omega_s.(d\alpha/dq)^2.T_2$$

where m is the reduced mass of the vibrating atoms and k is a lumped constant. Diamond has an exceptionally high Raman gain coefficient owing to both high $(d\alpha/dq)^2/m$ and T_2 values. There are several interesting characteristics of Raman lasers worth noting:

1) The equations for Raman amplification are closely analogous to conventional laser gain involving a population inversion. In the Raman case, the spontaneous Raman cross-section is analogous to the stimulated emission cross-section material parameter and the population inversion term is replaced by I_p.

2) Since gain is only present while a pump field is present, there is generally close temporal overlap between the output and pump pulses. As a result Raman lasers are often thought of as a nonlinear optical converters. Energy cannot be stored in the medium as in the case of a Q-switched population inversion lasers.

[†] For pump pulses of duration comparable or shorter than the phonon dephasing time T_2, the rate of accumulation of coherent phonons needs to be considered and the effective gain is reduced.

3) In contrast to nonlinear optical conversion process such as harmonic generation and four-wave mixing, Raman generation is automatically phase matched. That is, momentum is conserved in the interaction essentially independent of the momentum vectors of the pump and output beam. Momentum is conserved in the interaction since the scattered phonon in the Raman material carries away any recoil and consequently Raman lasers have several important properties. The phase properties of the Raman beam are constrained by design of the Raman resonator and as a result the spatial properties of the Raman output beam are often better than the pump, a property that enables Raman lasers to act as beam quality converters in a process often referred to as "Raman beam cleanup" [6]. This is unlike a phase-matched nonlinear conversion process where the phase properties of the output beam are directly related to that of the pump beam, an effect leads to exacerbation of distortions and hot-spots in the beam profile. Raman lasers can also be pumped at a range of angles non-collinear to the output beam axis such as in the side-pumping configuration often used in conventional lasers. A further corollary of automatic phase matching, is that the Raman process can be cascaded to generate an integral number of Stokes shifts. By careful Raman laser design, efficient generation at a selected Stokes order or at multiple Stokes orders can be achieved.

Raman laser designs can be divided into the two categories of external cavity and intracavity Raman lasers as shown in their most basic forms in Figure 1. Detailed reviews of design and performance of these systems are available in refs [7,8]. For external cavity Raman lasers, the cavity is designed to resonate the Stokes fields and the fundamental pump laser field is coupled into the resonator from a separate laser source. The input mirror is highly transmitting to the pump wavelength as practically possible, and the output mirror reflective at the pump wavelength to allow a double pass of the Raman medium. The spectral and spatial properties of the Raman output beam are dictated by the resonator design. An output Stokes order can be selected for example by designing the output mirror to output couple at the wanted Stokes order but reflect the lower Stokes orders. External Raman lasers operate most efficiently for pulsed pump lasers; however, continuous wave operation has been reported [9]. A major attraction of the external resonator is that it can be a simple add-on to an unmodified pump source, thus allowing the approach to leverage available laser systems as pump sources. Intracavity Raman lasers resonate both the pump and Stokes fields with the advantage of enhancing conversion in the Raman medium and enabling reduced pump power thresholds, and architecture well suited to compact diode-pumped devices capable of operating efficiently at low peak powers.

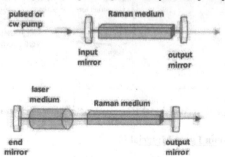

Figure 1 Basic Raman laser architectures: External cavity Raman laser (top) and intracavity Raman lasers (bottom)

Crystalline Raman materials offer the advantages of a solid-state material, rapid removal of waste heat (compared to gases and liquids), narrow Raman linewidths (compared to glasses) and high gain coefficients. Materials such as barium nitrate, potassium gadolinium tungstate, barium tungstate, yttrium vanadate and their close crystal relatives have been used widely. All these feature high gain coefficients and/or high damage thresholds that enable efficient Raman conversion to take place. Raman shifts are typically in the range 700-1332 cm^{-1} where diamond has the largest shift of all crystals widely used in Raman lasers. Low-order Stokes shifts thus allows access to important wavelength zones such as in the yellow – red, and the eye-safe region near 1.5 μm. Conversion efficiencies can be very high. For external cavity Raman lasers, for which it is straightforward to determine the conversion efficiency in the Raman medium, efficiencies greater than 50% are routinely observed [8]. Some crystals such the vanadates and the double metal tungstate also enable "self-Raman" laser action in which the medium acts as both the amplifier for the fundamental and Stokes fields. Due to the difficulty in doping diamond with active laser species the potential for diamond self-Raman lasers seems relatively limited.

The above discussion highlights the versatile properties of Raman lasers as optically pumped lasers for wavelength and beam quality conversion. A significant challenge that to date has limited integration of Raman lasers into applications is the weak nature of the Raman process (ie., the small Raman cross-section). As a consequence, high demands are placed on the spectral power density on the pump beam and the damage threshold of optical elements in order to create efficient devices. Transversely pumped Raman lasers are rarely done in practice as these requirements are even more difficult to satisfy. Improvements in pump lasers, optical coatings and Raman material quality over recent years have enabled the field to grow substantially and Raman lasers are finding numerous applications such is in ophthalmology [10], remote sensing [11] and astromonical guidestars [12]. The high gain of diamond along with its good resistance to optical damage is of significant interest for further increasing the performance range and utility of Raman lasers.

Physical	Chemical	Optical	Thermal	Electrical
Hard	Inert	High Raman gain	High thermal conductivity	High dielectric strength
High Young's modulus	Bio-compatible	Broad transmission	Low thermal expansion coeff.	
High tensile strength		Good damage resistance	**ALL HIGHLY VALUABLE FOR RAMAN LASERS**	
High speed of sound		High index		

Figure 2 Selected properties of diamond that are outstanding amongst other materials.

Diamond As a Raman Laser Material

Of all diamond's outstanding properties there are several that are immediately attractive to the Raman laser designer (Figure 2). The high Raman gain coefficient relaxes the constraints

274

on pump intensity, performance of optical coatings, parasitic losses and crystal length (the latter being fortunate currently since only relatively small crystals of high quality are available). The high thermal conductivity is promising for enabling Raman conversion at much higher average powers. The wide transmission range of diamond makes it of interest for generating wavelengths that fall outside the range of other materials at both the ultraviolet and infrared ends of the spectrum.

Table 1 contains a detailed comparison of the main parameters important to Raman laser design for diamond with other common Raman crystals. The thermal properties stand out most notably from the other materials. The thermal conductivity is over two orders of magnitude higher than the dielectric crystals, and 10-15 times higher than silicon. Since SRS deposits heat into the Raman material, this property is crucial for mitigating heat induced lensing and stress forces that introduce birefringence or lead to catastrophic damage. The outstandingly low thermal expansion coefficient of diamond also mitigates these problems. Though the thermo-optic coefficient (dn/dT) is relatively high, this will be counteracted by the rapid rate of heat removal and thus the moderation of temperature gradients. Most other Raman materials transmit in the range $0.35 - 5$ µm whereas diamond is also transparent in the ranges in the UV ($230 - 350$ nm) and at wavelengths longer than 6 µm. Note that significant absorption in diamond due to multi-phonon absorption occurs in the 3-6 µm range. The Raman linewidth, which is an indicator of the maximum line broadening introduced by Raman shifting, is at the low end compared to other materials but not as narrow as barium nitrate. Diamond is isotropic for linear optical phenomena, which is often considered a disadvantage because of the susceptibility for stress-induced birefringence to depolarize transmitted radiation. Stress-induced birefringence often inherent in CVD-diamond is discussed in more detail below. The laser damage threshold of diamond is also a crucial parameter, however, to date there is a lack of information available especially for the most recent material. Measurements on single crystal diamond suggest that the damage threshold is approximately 10 GW.cm^{-2} for pulsed 1064 nm radiation of duration 1 ns [13], and is probably higher than many other Raman materials.

	Crystal Class	Raman shift (cm⁻¹)	Line Width (cm⁻¹)	Stationary Raman gain @ 1µm (cm/GW)	Thermal Conductivity [W/m/K]	Thermal Expansion Coeff (x10⁻⁶ K⁻¹)	Thermo-optic coeff (dn/dT) [x10⁻⁶K⁻¹]	Transparency Range [µm]
LiIO₃	Uniaxial	822	5	4.8	-	-	-	0.38 - 5.5
KGd(WO₄)₂	Biaxial	768			2.5-3.8	2.5- 17	-1 - -5	0.3 - 5.0
		901	6	4				
Ba(NO₃)₂	Isotropic	1047	1	11	1.2	13	-20	0.3 - 1.8
BaWO₄	Uniaxial	926	1.6	8.5	3	11-35	-	0.4 - 3
GdVO₄	Uniaxial	884.5	3	4.5	5	-	4.7	0.3 - 2.5
YVO₄	Uniaxial	887.2	3.3	5	5	11	3	
Silicon	Isotropic	523	4.6	4	148	-	-	1.1 - 6.6
Diamond	Isotropic	1332.5	2	15- 20	>1800	1.0	20	0.23 - 3, >6

Table 1 Comparison of diamond's optical parameters with the most commonly used crystalline Raman materials.

EXPERIMENTS

Although diamond has long been known to be an interesting Raman laser material, it has only been the last few years in which Raman lasers have been demonstrated. In fact, not long after the discovery of the Raman effect by Raman and Krishnan in 1928, Ramaswamy discovered the strong and isolated 1332 cm^{-1} Raman mode in diamond [14]. Diamond was one of the first crystals that were used to exhibit SRS [15]. Though in principle Raman lasers made can be from natural diamond, indeed resonant effects in an uncoated natural diamond crystal were observed in 1970 [16], it has taken a reproducible supply of optical quality material provided by synthetic growth to enable any substantial diamond Raman laser development. Coinciding with the availability of low impurity CVD diamond in both the US and the UK in about 2005, reports were published for SRS in CVD-grown diamond [17] and for a micro-chip diamond Raman laser development [18] possibly also based on CVD diamond (details are available).

A program on Raman laser development at Macquarie University was initiated in 2008. An external cavity Raman laser based on a CVD diamond single crystal was designed to convert a standard 532 nm frequency doubled Nd:YAG laser to the first Stokes yellow wavelength at 573 nm [19]. Some optical properties of the Type IIa crystal (Element Six, UK) were previously characterized in ref [20] (refer Sample #6). The Raman laser performance was investigated for the arrangement in Figure 3. The crystal dimensions were 5 x 5 x 1.47 mm and it was oriented at Brewster's angle in order to avoid the use of anti-reflection coatings. Resonator mirrors were placed as close as practicable to the diamond chip, forming a total cavity length approximately 15 mm. The Raman laser was pumped with 10 ns pulses from the linearly polarized 532 nm laser (HyperYag, Lumonics) at the repetition rate of 10 Hz. The vertically polarized pump beam was compressed to approximately 0.35 mm beam diameter using a telescope.

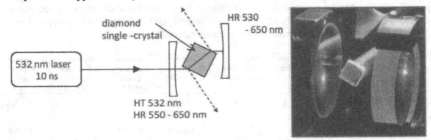

Figure 3 A schematic of the experimental layout for the 2008 Diamond Raman laser results [21].

Significant conversion to the first Stokes was observed for cavity mirrors highly reflective at the Stokes; however, essentially all of the Stokes energy was measured from the four reflections from the diamond facets. Though the device was not practical for applications, it enabled verification of lasing and determination of the efficiency of Raman conversion. By summing the output from each facet reflection, a maximum total conversion efficiency of 14% and slope efficiency 22% were derived. These values are markedly lower than for other materials (which as noted above is often >50%). By analysis of the polarization characteristics of the laser output and the expected facet reflection losses, it was determined that a major cause of the low efficiency was due to depolarization of the pump and Stokes fields on passage from the diamond. Poor preparation of the facets also caused additional losses and diminished the laser efficiency.

The main issues raised by this early study were used to develop an advanced version with highly efficient performance [22]. A new diamond crystal (Crystal 2) was manufactured (Element Six, UK) with dimensions 6.7 mm long, 3.0 mm wide and 1.2 mm thick using methods to reduce birefringence described in ref [19]. Figure 4 shows differential interference contrast images of the two crystals clearly showing the improved properties of Crystal 2 with respect to birefringence and surface preparation. The birefringence value for Crystal 2 is expected to be orders-of-magnitude lower than for Crystal 1.

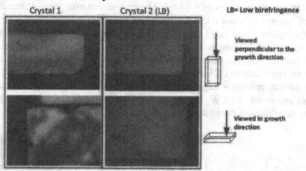

Figure 4 Comparison of differential interference contrast images for the two diamond Raman crystals. The images provide qualitative indication of the improved birefringence and surface preparation for Crystal 2.

Crystal 2 was Brewster cut for propagation along the longest crystal dimension and parallel to <011> and for propagation perpendicular to the growth direction to further minimize birefringence [19]. The Brewster facets were angled in the (011) plane so that the low-loss polarization was Raman-scattered with polarization parallel to the pump field. The third-order susceptibility tensor for diamond's crystal class [23] dictates preserved polarization in the (011) plane as confirmed by measurements of the polarized Raman spectra shown in Figure 5. As a result of this alignment, a slightly higher effective gain coefficient was accessed compared to the initial study in which the pump polarization was at a small angle (~23°) to the [011] direction.

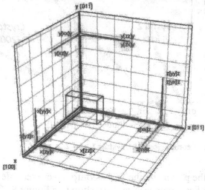

Figure 5 Backscattered polarized Raman spectra for a rectangular prism with {100} and {110} facets.

Performance was investigated using a similar external cavity resonator except for a larger output coupling at the first Stokes (29%T) and performance was investigated for a second pump laser at 5 kHz. The setup is shown in Figure 6. For the 5 kHz pump laser, input pump powers up to 2.2 W corresponding to pulse energies up to 0.4 mJ were used. Figure 7 shows the output energy as a function of the pulse energy incident on the crystal (factoring in the 5.8% input coupler reflection loss) compared against the Crystal 1 performance. The Raman laser threshold is approximately 0.1 mJ, above which the output increases linearly with slope efficiency 74.9% up to the maximum pulse energy of 0.24 mJ. The conversion efficiency at maximum energy is 63.5%. The efficiency is thus 4-5 times higher than for Crystal 1 and is similar, if not slightly higher than the highest efficiency values reported for external cavity Raman lasers using materials (viz., 64-65% for tungstate and nitrate materials as reviewed in ref [8]). The output largely consisted of first Stokes output at 573 nm. Second Stokes output at 620 nm was observed at high input energies (above 0.28 mJ), and increased to approximately 10% of the total Stokes output at the maximum input energy. In terms of output power the maximum combined first and second Stokes power was 1.18 W. Investigation of the performance at higher input powers was limited by the capability of the pump laser.

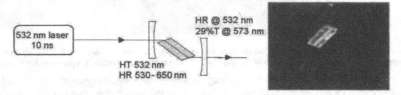

Figure 6 Schematic of experimental layout for the diamond laser operating with Crystal 2 [22].

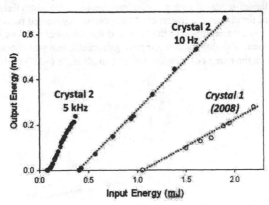

Figure 7 Output performance comparison of Crystals 1 and 2.

Pulse shapes of the pump and the Stokes output were recorded in order to analyze the temporal behavior of Stokes conversion and are shown in Figure 8. The onset of Raman lasing occurs when the pump power has attained approximately 30% of its peak value, causing a lag from the leading edge of the pump pulse of 1-2 ns. The calculated instantaneous conversion efficiency $I_s(t)/I_p(t)$ shows that the efficiency increases rapidly from zero to above 80% within 3 ns. The value peaks at 85%, which closely approaches the quantum efficiency ω_s/ω_p=92.8%. In terms of numbers of photons, more than 9 out of 10 pump photons are scattered into the Stokes beam at this time. After the peak, the conversion efficiency decreases steadily to approximately 40% when the pump intensity decreases to ~30% of its peak value. At longer times ($t > 15$ ns) detector saturation for low signals leads to large errors in the calculated values and are not shown. Figure 8 also shows the pulse shape of the depleted pump beam after making the double pass of the Raman laser. The depleted pulse behavior prior to the onset of Stokes conversion closely matches the pump pulse as expected. Once threshold is attained ($t > 3.5$ ns), large depletion occurs as revealed by the rapid decrease in the depleted pulse intensity while the pump increases. It is deduced that the balance between pump and output energy is accounted for by unconverted pump photons during all stages of the pump pulse (ie., prior, during and after the Stokes pulse). Though there is measurable pump absorption (<1.1 cm^{-1} at 532 nm as obtained by calorimetric measurements), the pulse shapes indicate it does not significantly impact the conversion efficiency under these conditions.

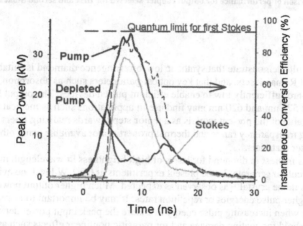

Figure 8 Pulse shapes of the incident pump, the Stokes output and the residual pump beam after double passage through the crystal [22].

For the 10 Hz pump laser, output pulse energies up to 0.64 mJ were achieved. The efficiency was lower than for the 5 kHz laser as we used a pump spot in the diamond crystal larger than the resonator mode in order to ensure we operated safely away from the damage threshold. It should be noted however that we observed no damage to Crystal 2 when either using the 10 Hz or 5 kHz pump lasers at incident pump intensities up to approximately 0.5 GW/cm^2. A comparison of performance using output couplers suited for first and second Stokes generation (see Figure 9) demonstrated slightly lower efficiency for the 620 nm second-Stokes performance (slope efficiency 48% cf., 64% for similar conditions using a first Stokes output coupler).

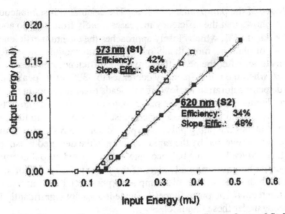

Figure 9 Comparison of performance for output coupler selected for first and second Stokes output (10 Hz pump laser).

DISCUSSION

The results demonstrate that synthetic low-birefringence diamond is suitable for realizing highly efficient Raman lasers, and that key optical parameters such as absorption, scatter and depolarization are sufficiently low to enable efficient pulsed devices. The output laser wavelengths at 573 nm and 620 nm may find use in applications such as medical and biosensing; however, the value of the present work is as a major step towards realizing lasers that leverage the outstanding transparency range and thermal properties not available using other Raman and non-Raman laser materials.

There is interest in diamond for lasers of high brightness in wavelength in regions otherwise difficult to generate. In the present experiments at the 1 W level, no evidence for thermal effects in the crystal was observed as expected. Much higher output powers are likely by using either higher pulse energies or repetition rates. It may be important to increase the beam waist diameter when increasing pulse energy to ensure the peak input power densities remain below the threshold for coating damage and for parasitic nonlinear effects such as self-focusing. On the simple basis of the diamond's high thermal conductivity, thermal lensing effects are not expected for Stokes powers approximately two orders higher than other Raman materials. Given that current output powers are currently approaching 10 W, there is promise for diamond to scale to multi-hundred watt diamond Raman lasers without performance impacted by thermal lensing (though the isotropic nature of diamond will require consideration of thermally induced stress birefringence).

Figure 10 shows a comparison of diamond's transparency range against other key Raman materials. The ultraviolet range between 270-320 nm, where diamond has a distinct advantage, is an interesting region for applications in sensing, defence and materials processing. For wavelengths longer than 6 μm there is·a paucity of alternative materials yet strong demand for source in trace gas sensing, security and defence. There are significant challenges for long wavelength extension, however, due to the presence of diamond's multi-phonon absorption band between 3-6 µm and also the diminishing gain coefficient for lower ω_s.

Diamond Raman laser development is very much in its infancy and the outlook for rapid growth in the field is excellent. New high performing devices are likely to emerge quickly given the high efficiency already obtained for external cavity diamond Raman lasers and aided by the large and existing knowledge base in intracavity Raman lasers, and in configurations that address Raman beam cleanup, 'wavelength switchablility', and multi-wavelength and continuous wave operation [5-7].

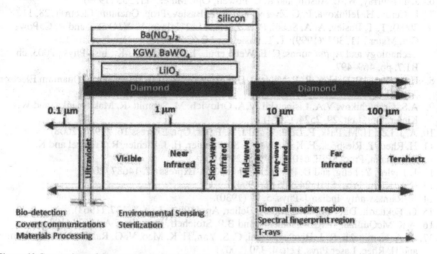

Figure 10 Comparison of diamond transparency range with other representative Raman laser materials.

CONCLUSIONS

1)   The high Raman gain, high thermal conductivity and broad transmission of diamond are highly attractive properties for Raman lasers with capabilities that extend well beyond those possible using alternative materials.
2)   Highly efficient diamond Raman lasers operating in a pulsed nanosecond regime have been demonstrated using a Brewster cut crystal of high quality and low birefringence diamond grown by CVD.
3)   Prospects are excellent for demonstrating diamond Raman lasers of high average power and in wavelength ranges not well serviced by other materials within the next few years.

ACKNOWLEDGMENTS

The author thanks in particular Jim Rabeau, Jim Butler, Ian Friel and Alex Sabella for their valuable contributions to this work. This work was supported by the Macquarie University Vice-Chancellor's Innovation Fellowship Scheme 2009-2011.

REFERENCES

1. S.C. Rand and L.G. DeShazer, Opt. Lett. 10, 481 (1985).
2. P. John, Science 292, 1847-1848 (2001).
3. Patent: K. Jamison and H. Schmidt, "Doped diamond laser," US Patent No. 5,504,767 (1996).
4. A. Penzkofer, A. Laubereau, and W. Kaiser, Progress in Quantum Electronics 6, 55-140 (1979).
5. H. M. Pask, Prog. Quantum Electron. 27, 3 (2003).
6. J. T. Murray, W. L. Austin, and R. C. Powell, Opt. Mater. 11, 353 (1999).
7. P. Cerny, H. Jelinkova, P. G. Zverev, and T. T. Basiev, Prog. Quantum Electron. 28, 11 (2004); T. T. Basiev, A. A. Sobol, P. G. Zverev, L. I. Ivleva, V. V. Osiko, and R. C. Powell, Opt. Mater., 11, 307, (1999); T. T. Basiev and R. C. Powell, in Handbook of Laser Technology and Applications, C. E. Webb et al., Ed. London, U.K.: Inst. Phys., 2003, ch. B1.7, pp. 469–497.
8. H. M. Pask, P. Dekker, R. P. Mildren, D. J. Spence, and J. A. Piper, Prog. Quantum Electron. 32, 121 (2008).
9. A.S. Grabtchikov, V.A. Lisinetskii, V.A. Orlovich, M. Schmitt, R. Maksimenka, and W. Kiefer, Opt. Lett. 29, 2524, (2004).
10. A. J. Lee, H. M. Pask, P. Dekker, and J. A. Piper, Opt. Express 16, 21958 (2008).
11. H. Rhee, T. Riesbeck, F. Kallmeyer, S. Strohmaier, H. J. Eichler, R. Treichel and K. Petermann, Proc. SPIE, 6103, 610308 (2006).
12. L.Taylor, Y. Feng, and D. Bonaccini Calia, Opt. Express 17, 14687 (2009).
13. C.A. Klein, Proc. SPIE 2428, 517 (1994)
14. C. Ramaswamy, Indian J. Phys. 5, 97 (1930).
15. G. Eckhardt, D. P. Bortfeld, and M. Geller, Appl. Phys. Lett. 3, 137, (1963).
16. A.K. McQuillan, W.R.L. Clements, and B.P. Stoicheff, Phys. Rev. A 1, 628 (1970).
17. A. A. Kaminskii, R. J. Hemley, J. Lai, C. S. Yan, H. K. Mao, V. G. Ralchenko, H. J. Eichler, and H. Rhee, Laser Phys. Lett. 4, 350 (2007).
18. A.A. Demidovich, A.S. Grabtchikov, V.A Orlovich, M.B Danailov and W. Kiefer, in Conf. Dig. Lasers and Electro-Optics Europe, (Optical Society of America, Washington DC, 2005) pp. 251.
19. I. Friel, S.L. Clewes, H.K. Dhillon, N. Perkins, D.J. Twitchen, G.A. Scarsbrook, Diamond and Related Materials, 18, 808-815, (2009)Journal article: 1. S.W. Bonner and P. Wynblatt, J. Mater. Res. 1, 646 (1986).
20. G. Turri, Y. Chen, M. Bass, D. Orchard, J.E. Butler, S. Magana, T. Feygelson, D. Thiel, K. Fourspring, R.V. Dewees, J.M. Bennett, S. Pentony, S. Hawkins, M. Baronowski, A. Guenthner, M.D. Seltzer, D.C. Harris and C.M. Stickley, Opt. Eng. 46, 064002, (2007).
21. R.P. Mildren, J.R. Rabeau, J.E. Butler, *Opt. Express*, 16, 18950 (2008) .
22. R.P. Mildren and A. Sabella, Opt. Lett. 34, 2811 (2009).
23. D.J. Gardiner, P.R. Graves, H.J. Bowley, *Practical Raman Spectroscopy*, (Springer-Verlag, 1989) p. 24

AUTHOR INDEX

SUBJECT INDEX